Optimizing Voice in ATM/IP Mobile Networks

Optimizing Voice in ATM/IP Mobile Networks

Juliet Bates

McGraw-Hill
New York Chicago San Francisco Lisbon
London Madrid Mexico City Milan New Delhi
San Juan Seoul Singapore Sydney Toronto

Cataloging-in-Publication Data is on file with the Library of Congress

McGraw-Hill

A Division of The McGraw-Hill Companies

1 2 3 4 5 6 7 8 9 0 DOC/DOC 0 9 8 7 6 5 4 3 2

ISBN 0-07-139594-6

The sponsoring editor for this book was Marjorie Spencer and the production supervisor was Pamela A. Pelton. It was set in Century Schoolbook by MacAllister Publishing Services, LLC.

Printed and bound by R.R. Donnelley & Sons Company.

This book is printed on recycled, acid-free paper containing a minimum of 50 percent recycled de-inked fiber.

To my husband Ted

Contents

Foreword

For years, the transmission of voice over the PSTN has been implemented over TDM networks. Traffic engineering rules for these networks have been understood for well over half a century, but the shift to voice over packet networking introduces a different operating paradigm. The challenge to network operators is to deliver a *Quality of Service* (QoS) that at least matches the voice quality that TDM networks have achieved, while reducing costs through improved bandwidth efficiency and statistical gains.

The TDM codec G.711 is inefficient with respect to the use of network resources. Complex voice codecs, such as G.729A/B, or mobile codecs such as AMR, ACELP, or FR-GSM can encode voice signals into data units that produce throughput rates at far less than the G.711 rate of 64 Kbps.

Furthermore, conversational voice involves two states, that of talking and that of listening. When a person is listening, there is no need to send digitized voice samples into a packet or cell relay network towards the active talker. This concept is known as silence suppression, voice activity detection, or discontinuous transmission and lends well to achieving statistical gains in a packet network.

The purpose of this book is to provide an insight into some of the challenges and decisions facing network engineers and planners as they deploy voice services over their packet networks. The objective is to explain the dynamics of conversational speech and to empower the network planner with the requisite knowledge to accurately size the traffic descriptors of multiplexed voice channels.

To that end, the author, Juliet Bates, seeks to define a representative model of voice traffic in order to make recommendations for the effective use of the sophisticated traffic management capabilities of ATM. This book describes how to determine the extent to which overbooking can be applied in order to achieve statistical gain of silence suppressed and compressed voice channels, while providing a *Cell Loss Ratio* (CLR) guarantee.

Taking this a step further, it is proposed in this book that the determination of the *Peak Cell Rate* (PCR) , *Sustainable Cell Rate* (SCR) , and *Maximum Burst Size* (MBS) ATM traffic descriptors should be based on the mathematical modeling of conversational speech.

Networks should be, and *can* be, engineered such that overprovisioning is not necessary. ATM mechanisms for QoS are suitable for voice services, despite the fact that conversational behavior is memoryless (statistically speaking).

This book describes how to maintain QoS for voice traffic. ATM already has the requisite QoS capabilities. Some of the principals are applicable to IP networks, but to truly engineer QoS, service providers must look to ATM. Readers of this book will learn about the opportunities for statistical gain, what is practical, and what compromises or risks can be taken to achieve a high QoS at optimal network utilization.

Brian A. Day
Product Line Manager
Voice Gateways
Broadband Networking Division
Alcatel Canada

Preface

Economic gains in voice transmission over a network are made possible by two factors: low bit rate through compression, and exploitation of speech silence durations. Overestimation of bandwidth requirements will cause underutilization of the network resources and increased call blocking probability. Underestimation may cause cell or packet loss. The potential for statistical gain in multiplexed voice sources is inhibited by the intermittent use of single voice channels for fax and uncompressed voice, which may complicate the normal methods of setting traffic descriptors. Overbooking can be applied to compensate. The extent to which overbooking can be applied is a subject of this book together with calculations to show the risk of cell loss and the associated delay (and delay variation) incurred in the buffer.

The *Quality of Service* (QoS) parameters, *Cell Loss Ratio* (CLR) and *Cell Delay Variation* (CDV), which apply to *Asynchronous Transfer Mode* (ATM), *Constant Bit Rate* (CBR), and *Variable Bit Rate* (VBR) service classes are controlled by the size of the buffer at the egress port and the service rate of the queue in the buffer. The target CLR will reflect the probability that the buffer could be exceeded. In provisioning effective bandwidth, the *Call Admission Control* (CAC) Algorithm is aware of the QoS targets required by the selected service class. This book describes a method for the reduction of bandwidth, due to silence suppression, and the determination of a scaling factor which can be applied with a risk of cell loss selected in order to maintain the CLR and CDV of the chosen ATM service category.

Juliet Bates
Principal Consultant
Professional Services Group
Broadband Network Division
Alcatel Telecom UK

Acknowledgments

I wish to acknowledge Brunel University, Uxbridge, UK for allowing me to reproduce material from my Ph.D. thesis (July 2001).

Also, I particularly thank Dr. Michael Berwick in the Department of Electronic and Computer Engineering at Brunel, and Professor Malcolm Irving, who both provided excellent support and advice during my time at Brunel University.

I gratefully acknowledge the technical advice I have received from Mustapha Aissaoui.

I also thank Bryan Edwards for his encouragement, and Tim Forshaw, Dave Hills, and Mike Wilkinson for kindly reviewing each draft and providing helpful comments and advice.

Within the Appendixes of this book, there are extracts from comprehensive reports into the market development prospects for voice over alternative technologies, and I thank the Yankee Group for permitting this material to be included.

Introduction

Familiarity with the fundamentals of *Asynchronous Transfer Mode* (ATM) networking is assumed. While a very large number of books have been published on this subject, Bibliography references [1] and [2] will provide a grounding in the areas covered in this work.

Chapter 1 provides an introduction and describes voice access models for mobile voice networks, such as the *Universal Mobile Telecommunications System* (UMTS), combining the ATM and *Internet Protocol* (IP) in the core.

Chapter 2 describes the voice access methods and signaling protocols employed in fixed voice networks—for example, where a *Private Branch Exchange* (PBX) accesses an ATM core network.

Chapter 3 discusses recently completed research in voice characterization and defends the rationale for choosing the distribution and parameters best for a voice model. Also included are alternative ways of building a model that will approximate the arrival process using key characteristics obtained from real traffic.

Chapter 4 describes the framing format when voice is transmitted over an ATM network. Three different types of *ATM Adaptation Layer* (AAL) can be used to encapsulate coded voice. This section helps you assess the framing overheads incurred by each type of AAL.

Chapter 5 explores various methods of servicing the contributing channels used by each AAL type and calculates the corresponding delay implied by each solution. The difficulties of sizing bandwidth requirements for compressed and silence-suppressed voice channels in the presence of other applications are explained.

Chapter 6 sets out the *Cell Loss Ratio* (CLR) and *Cell Delay Variation* (CDV) performance targets, which must be achieved to ensure adherence to a particular ATM *Quality of Service* (QoS)

Class. Methods of estimating the traffic descriptors *Peak Cell Rate* (PCR), *Sustainable Cell Rate* (SCR), and *Maximum Burst Size* (MBS) for silence-suppressed voice channels are outlined, and the risks involved in building a representative voice model are weighed.

Chapter 7 further defines the voice model at talkspurt level and at cell level, and establishes the mathematical theory used to discover the bandwidth requirements of silence-suppressed voice channels in the presence of an estimated and quantified risk of cell loss.

Chapter 8 discusses the sizing of a burst according to the number of sources likely to continue in a talkspurt in order to show how much the traffic descriptors can be reduced without affecting the QoS parameters of the selected traffic class.

Chapter 9 applies a *Call Admission Control* (CAC) Algorithm. Effective bandwidth is compared before and after the traffic descriptors have been reduced according to the probability tables described in this book.

Chapter 10 records the results of Opnet simulations of an M/M/1 queue. The simulations are used to confirm the arrival behavior of exponentially-sized talkspurts into a queue served at an exponential rate.

Chapter 11 presents the main conclusions of the book and the research that informed this expedition into the future.

Chapter

1

Mobile Voice Access to ATM/IP Networks

The world of communications today is experiencing a major trend toward mobility on the one hand and toward an ever-increasing usage of the Internet on the other hand. Driven by these influences, it is envisaged that mobile communication and data communication will merge, and today's *Global System for Mobile Communication* (GSM) networks, which are optimized for voice communication, will evolve toward multimedia mobile networks.

This concept is referred to in Europe as the *Universal Mobile Telecommunications System* (UMTS). UMTS was initially developed within the *European Telecommunications Standardization Institute* (ETSI). UMTS is being standardized by the *Third-Generation Partnership Program* (3GPP) working in collaboration with the *International Mobile Telecommunications 2000* (IMT 2000) project. IMT 2000 is an initiative of the *International Telecommunications Union* (ITU) to provide wireless access to the global telecommunication infrastructure, through both satellite and terrestrial systems, serving fixed and mobile users in public and private networks. UMTS is shown in Figure 1-1.

Current GSM networks support voice and low-speed data services that are circuit switched—that is, the traffic is carried between users in bearer circuits that are switched under the control of signaling from the users.

The starting point for the development of the *Third-Generation* (3G) UMTS architecture is the introduction of a modified Internet protocol, *General Packet Radio Service* (GPRS), preparing the circuit-switched GSM networks for packet switching. While GPRS

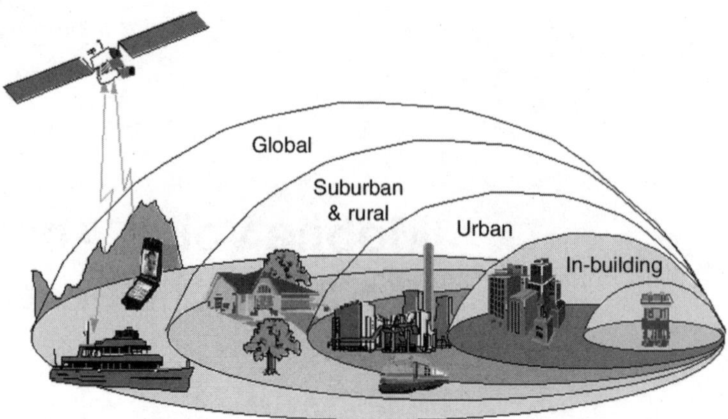

Figure 1-1 The UMTS

will provide mobility, and a convenient and efficient way to transport data, it will also build a bridge toward the introduction of a mobile Internet.

1.1 Second- and Third-Generation Mobile Networks Standards

Today's *Second-Generation* (2G) mobile networks follow different standards in different regions of the world. The European GSM standard is effectively a de facto world standard currently claiming around 60 percent of the total mobile telephone population. The United States and Japan have different standards such as the *Digital Advanced Mobile Phone Service* (D-AMPS), *Code Division Multiple Access* (CDMA), and *Personal Handyphone System* (PHS). The *Third-Generation* (3G) of mobile communications will have a more common approach. The 3GPP includes participants from the United States, Europe, and Asia.

The standardization of UMTS is being carried out by the ITU-T SG11 and ETSI *Special Mobile Group* (SMG) and *Network Aspects* (NA). UMTS Release 1999 is the first phase of the 3GPP standards for UMTS. This is a completed set of standards that define a UMTS network able to provide users with voice and data services in such a way as to be fully compatible with the present GSM and GPRS systems.

UMTS Release 2000 has been divided into Release 4 and Release 5 where the transport and control planes are separately implemented, giving the flexibility to equally use the *Asynchronous*

Transfer Mode (ATM) or *Internet Protocol* (IP) as transport networks.

1.2 ATM and IP Transport in UMTS

As shown in Figure 1-2, two distinct domains currently exist within UMTS Release 1999:

- **The circuit-switched domain** Based on GSM, this is used for connections to the *Public Switched Telephone Network/ Integrated Services Digital Network* (PSTN/ISDN), mainly for real-time applications (for example, speech).
- **The packet-switched domain** For connections to external IP networks, this uses a standardized overlay GPRS network supported by the *Serving GPRS Support Node* (SGSN) and the *Gateway GPRS Support Node* (GGSN).

The *UMTS Terrestrial Radio Access Network* (UTRAN) links to the core network via a *Radio Network Controller* (RNC), which controls the interface between the radio access and the ATM/IP transport in the core. A *Mobile Switching Center* (MSC) connects the circuit-switched domain to the PSTN via an ATM or a *Time-Division Multiplexing* (TDM) network. An IP gateway receives packetized data, which is carried via the IP transport plane.

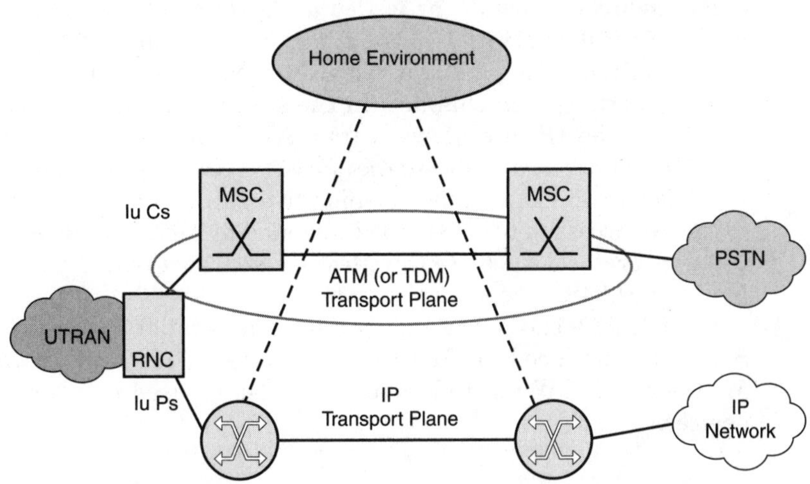

Figure 1-2 ATM and IP transport planes in UMTS Release 1999

1.3 ATM in UMTS

In UMTS Release 1999, ATM can be used as the transport technology:

- Between the radio *Base Transceiver Station* (BTS) and the RNC
- Between RNCs
- In the Iu interface between the RNC and the core network
- The Iu interface separates into
  - The Iu-CS interface between the RNC and the core network circuit-switched domain
  - The Iu-PS interface between the RNC and the core network packet-switched domain

Both circuit- and packet-switched services are carried in ATM cells. An *ATM Adaptation Layer* (AAL) protocol must be selected appropriately. In the case of voice transmission, this will normally be AAL2 for voice bearer traffic and AAL5 for signaling.

Within the core of the network, voice services may be transmitted either on circuit-switched TDM (as in the existing GSM network) or via ATM transport. If ATM is used, AAL2 would be the recommended adaptation layer, but AAL1, AAL5, or a *Voice over IP* (VoIP) solution also could be used.

In Release 4 of UMTS, the Iu-CS interface and any inherited GSM 'A' interfaces are terminated at the point of entry to the circuit-switched domain by a *Media Gateway* (MGW). The network service delivery and control functions are separated from the voice traffic. ATM signaling is relayed to the MSC call server over IP transport. Speech channels in the Iu-CS interface can be relayed into the IP core network from ATM AAL2 transport onto *User Datagram Protocol / Internet Protocol* (UDP/IP) transport. A MGW, operating in a similar fashion to a gateway currently used to control access to the PSTN, is controlled by an MSC call server, using a protocol such as the *Media Gateway Control* protocol MeGaCo/H.248.

The IETF MeGaCo Working Group, in cooperation with the ITU SG16 has produced the MeGaCo protocol, which is used between MGWs and MGW controllers and can also be used to control IP devices.

1.4 The UTRAN

The basic building blocks of the UTRAN are the Node B and the RNC. The Node B is approximately equivalent to the GSM BTS in that it links an antenna site to the network. Most Node Bs will initially support E1/T1 interfaces. In the 3G network, an RNC is often colocated at the GSM *Base Station Controller* (BSC). However, an RNC will generally provide considerably more capacity to link in Node Bs than a BSC. The transport network between the BTS/Node B and the BSC/RNC may be ATM, IP, or a *Synchronous Digital Hierarchy* (SDH) ring.

At the BSC/RNC, the traffic divides into voice and data. Packetized data travels to the IP backbone via an SGSN and a GGSN. The UMTS architecture, therefore, allows for two main paths through the network:

- One path for native IP traffic (through the SGSN, GGSN, Multimedia Call Server, and gateways)
- One path for non-IP traffic (through the MSC server and associated gateways) guaranteeing *Quality of Service* (QoS) for real-time applications such as voice

Figure 1-3 shows multiservice access in the UTRAN.

1.5 Speech Coding in the UTRAN

In UMTS Release 1999, speech is sampled at the source by the user's equipment and then coded for transmission. *Adaptive*

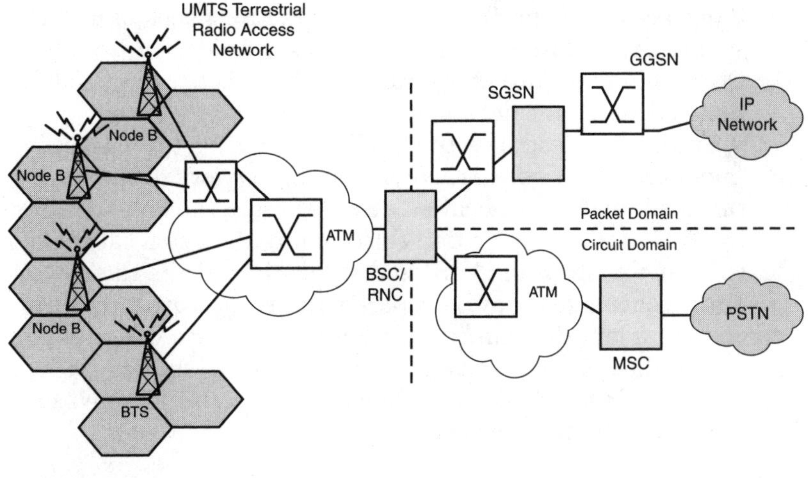

Figure 1-3 Multiservice access in the UTRAN

Multirate (AMR) coding is the default method of speech coding, and this must be supported by all mobile user equipment. Other speech coding may be optionally selected.

Transcoding from AMR (or any other speech coding) to G.711 is performed in the MSC. However, if the user's equipment at both ends of a voice call uses the same coding, then transcoding to G.711 is not necessary.

1.6 QoS in IP Networks

The Yankee Group [76] forecasts that high-end voice applications will migrate in favor of ATM rather than in favor of the Internet, even though there are indications of a belief that the public Internet as a whole will become more reliable and secure to the point where it will eventually deliver acceptable quality voice.[1] Network management in a private IP network can ensure the provision of much better voice quality and availability than the public Internet.

The *Internet Engineering Task Force* (IETF) *Real-Time Transport Protocol* (RTP) is used to synchronize the voice samples and control the sampling rate of the voice samples, which are transported in UDP/IP packets across the GPRS and core IP networks.

The user's equipment must activate a bidirectional *Packet Data Protocol* (PDP) context between itself and the called user's equipment. The establishment of the PDP context assigns an IP address and bandwidth for signaling.

PDP uses GPRS to allocate the bandwidth and QoS needed for the speech packets. To meet the interactive needs of multimedia services, the GPRS network will have to provide increased levels of QoS. As in fixed VoIP networks, speech packets can take different routes through the network.

Swale [79] describes the challenges facing the packet-based, "media-neutral" infrastructure of an IP network designed to meet the needs of data applications. Swale presents results of technical trials carried out by British Telecom into the technical challenges currently confronting VoIP and states that "while the benefits of VoIP technologies may be apparent, there are currently a number of fundamental technical challenges that remain to be solved within the industry. These mainly center upon the provision of applications in anything other than a purely best-effort environment —specifically the issues of

[1] Appendix E includes further extracts from reports by the Yankee Group.

- Achieving manageable QoS across IP networks
- Resolving complex interactions at the myriad of boundary points in an IP network
- Achieving appropriate network and service management functionality."

Project *Telecommunications and Internet Protocol Harmonization over Networks* (TIPHON) is a project led by *European Telecommunications Standards Institute* (ETSI). TIPHON has eight working groups tackling areas of VoIP, including requirements for service interoperability, architecture, call control, and QoS, as well as mobility and security.

3G.IP has developed an initial architecture for 3G using an IP call server approach and the H.323 protocol and *Session Initiation Protocol* (SIP). IP links are established to call servers. The call server instructs the mobile to set up an IP bearer channel to a gateway and then instructs the gateway to set up a VoIP channel to the PSTN.

Work within the ITU-T is looking into how SIP can be extended to support call control across different networks and allow for compression of the signaling messages it carries.

The *Differentiated Services* (Diffserv) framework (RFC 2474 and RFC 2475) has been proposed within the IETF to provide multiple QoS classes over IP networks. See Bibliography [3], [4], and [5].

IP version 6 (RFC 2460) offers larger address space and support for prioritization of packets, and it is planned that Release 5 of UMTS will be based on an IP version 6 core network.

1.7 Media Gateways (MGWs)

Interworking between today's legacy network architecture and the next generation UMTS 2000 architecture is facilitated by an MGW. Gateways provide a number of interworking functions from high to low end systems, ranging from single-line IP telephony adapters to MGWs, supporting thousands of simultaneous calls between two networks.

A MGW terminates bearer channels from a switched circuit network (that is, 64-Kbps channels) and media streams from a packet network (for example, RTP streams in an IP network).

The following functions are performed by an MGW:

- Interaction with a *Media Gateway Control Function* (MGCF).
- Management of resources such as echo cancellation, transcoders, conference bridges, and so on.

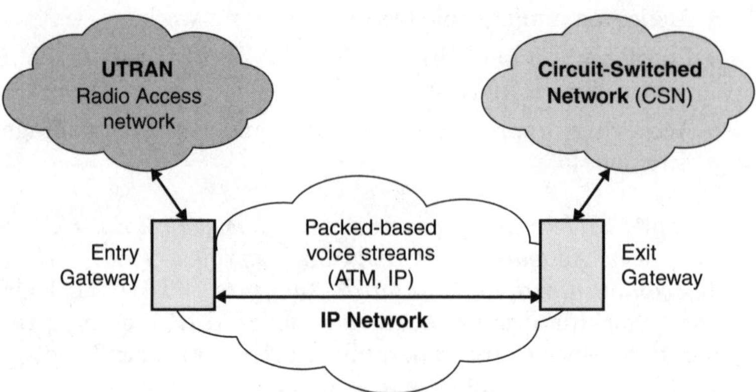

Figure 1-4 An example of entry and exit gateways

- Processing of audio, video and data, and full-duplex media translation. This involves having support for codecs such as G.711, G.722, G.723.1, G.728, G.729, H. 261, and T.120.

Figure 1-4 shows an example of entry and exit gateways enabling a seamless translation of information between a UMTS network, with a combined ATM/IP core, handing VoIP packets to the legacy circuit-switched telephone network.

The entry gateway shown in Figure 1-4 is located at an RNC at the border of the IP core network. There are two types of RNC. An RNC may provide AAL2/ATM in the user plane and SCCP/MTP3b/SAAL-NNI/ATM in the control plane, connected to a full-IP core network, in which case the MGW performs bearer conversion for the user plane to the core—for example, AAL2 toward RTP/UDP/IP. The AAL2/ATM media bearer (Iu) is terminated here. There is no transcoding and the media flows are transported transparently through the gateway (an exception may be a conference call). Alternatively, the RNC may be IP, in which case no bearer conversion is required.

The exit gateway is located at the border of the UMTS core network and the PSTN. This MGW is used as a termination point for the UMTS all-IP core network when voice traffic is being transported to the PSTN. In this gateway, media processing takes place, such as transcoding and echo cancellation. Toward the PSTN, the MGW will perform the handling of AMR/RTP/UDP/IP to G.711/G.704 and will also act as a firewall.

Figure 1-5 shows an example of the protocols involved. In this example, there is a speech connection from a mobile telephone to

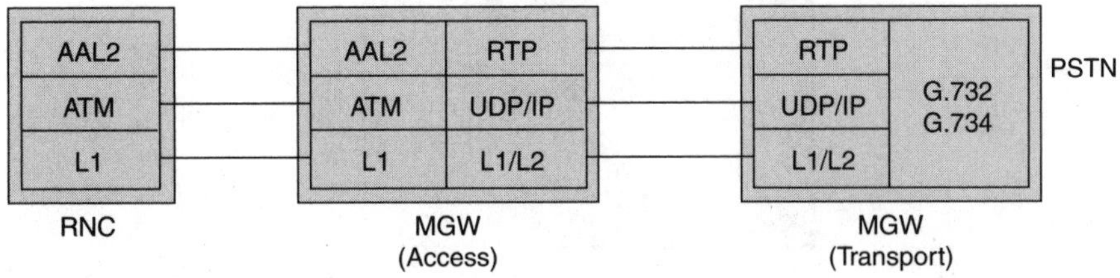

Figure 1-5 Protocols involved in the transport of mobile voice to the PSTN across an IP core

a terminal in the PSTN network. The mobile uses the AMR codec to code the speech. The speech frames in this example are transported over AAL2/ATM. The AAL2 connection is terminated in the MGW, and the speech frames are transported via RTP/UDP/IP over the all-IP core network to the MGW at the border to the PSTN. There the speech frames are transcoded to a G.711 codec and forwarded to the PSTN. Figure 1-5 illustrates the protocols involved.

1.8 Media Gateway Controller

MGWs will be controlled by a *Media Gateway Control Function* (MGCF) such as is provided by the MeGaCo/H.248 protocols, as illustrated in Figure 1-6. MeGaCo/H.248, which complement H.323 and SIP, are cost-reducing standards intended to divide a gateway into several separate functional entities so that as new features are added, it doesn't mean that every box in the network has to be updated. Further tailoring of the H.248 protocol will be required in order to support additional codecs such as AMR.

The H.323 standard is an ITU standard that enables voice, data, and video to be sent over an IP network. The IETF developed the SIP. H.323, and SIP are competing standards in that they both deal with passing voice, video, and multimedia over IP networks. H.323 has been a standard since 1996. At the time of this writing, SIP is still a draft. H.323 systems can also make use of the IETF protocols *Lightweight Directory Access Protocol* (LDAP) and *Resource Reservation Protocol* (RSVP).

A further function of a MGW is support for mobility. The MGW bearer control and the payload-processing capabilities will also need to support mobile-specific functions such as the *Serving Radio Network Subsystem* (SRNS), relocation/handover and

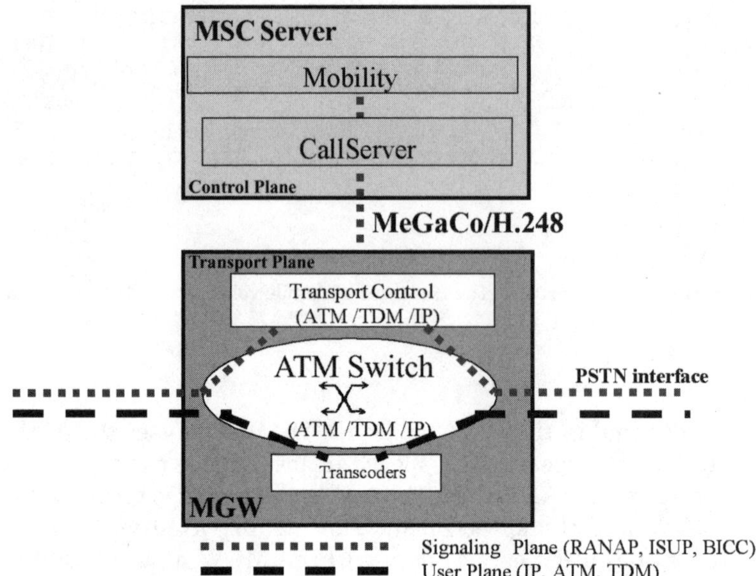

Figure 1-6 Control of the media gateway

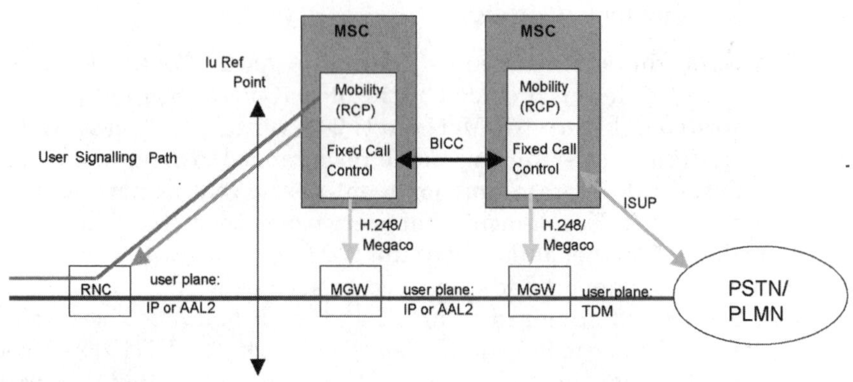

Figure 1-7 MSC control of the MGW

anchoring (GPRS). It is expected that current H.248 standard mechanisms will be applied to enable this.

Figure 1-7 illustrates how MSCs will handle call control using the *Bearer Independent Call Control* (BICC) protocol. The *ISDN User Part* (ISUP) protocol is used to carry signaling between an MSC and the PSTN/*Public Land Mobile Network* (PLMN).

1.9 Interworking Between IP and ATM

When IP traffic from many sources is generally carried over the same ATM virtual connection, the aggregate traffic behavior is difficult to characterize. Traffic shaping becomes essential.

Gibbens and Kelly [40], [41] have looked at distributed connection admission control with an edge device or broker acting as a gateway to determine whether to accept a connection or not. Kelly, Key, and Zachary [46] looked at a packet-marking system used by an end system or user to determine whether a connection should enter the network.

In summary, IP is moving forward toward supporting QoS assurances and is therefore designing traffic-management concepts similar to those defined for ATM (for example, policing, CAC, and flow control). Interworking is being developed to map between QoS classes and traffic parameters in IP and ATM.

Chapter

2

PBX Voice Access to ATM

2.1 Disadvantages of Inherited PBX Solutions

In a typical *Private Branch Exchange* (PBX) implementation, call routes must be configured in every PBX. The primary and alternate routes specified for each possible destination are sized to take into account the peak traffic volume bandwidth required on each route as well as the required grade of service. There are many costs and problems associated with this approach:

- **High operational costs** High operational costs are incurred in order to maintain PBX routing tables and reconfigure network bandwidth when traffic patterns or volumes change.
- **Full meshing** For full connectivity, either all PBX nodes have to be fully meshed with point-to-point private circuits, or (more likely) they must be interconnected via a tandem switch.

  A large corporate network typically contains multiple "tandem" central office switches, or PBXs, between the point of origin and the call's final destination.
- **Unnecessarily blocked calls** Unnecessarily blocked calls, particularly in complex network topologies, can result in an overflow to *Public Switched Telephone Network* (PSTN) links.
- **Inefficient bandwidth usage** Bandwidth usage is inefficient due to dedicated bandwidth allocations.
- **Quality degradation** Quality degradation can be caused by compression techniques in multiple hop configurations. Because the compression algorithms used approximate the signal closely,

but not exactly, the signal quality is degraded with each additional compression/decompression. The cumulative effect of many hops can be poor-quality speech.

2.2 Advantages of Migrating Voice to ATM

The advantages of migrating voice traffic to an *Asynchronous Transfer Mode* (ATM) backbone are as follows:

- The ability to combine data traffic and voice over the same connections and the ability to deliver multiple *Quality of Service* (QoS) options for voice traffic
- A reduction in the PBX interface hardware requirements because the number of egress interfaces required will be dictated by the total bandwidth required rather than by the number of multiple trunk groups needed in order to fully intermesh all remote PBXs
- A reduction in call charges (to local rate) through toll bypass

A key aim in building an Enterprise network is the ability to integrate multiple traffic types from multiple locations over a common backbone using a single access technology. Figure 2-1 represents a typical Enterprise solution for integrated voice and data networking.

Voice access is provided to an Enterprise PBX over narrowband-leased lines. Signaling is *Channel Associated Signaling* (CAS) or *Common Channel Signaling* (CCS).

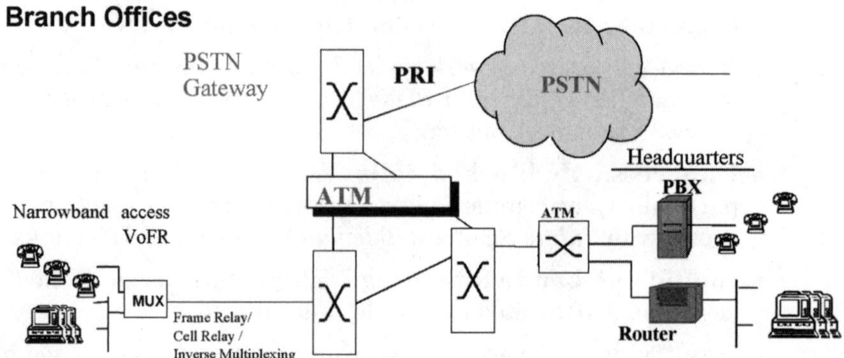

Figure 2-1 Narrowband access to ATM

LAN traffic, typically IP-based from routers, can be integrated with the voice traffic, over a frame relay (or cell relay) access link, sharing a common ATM backbone.

As shown in Figure 2-1, an ATM linecard providing voiceband services in an ATM switch is connected to a central office switch by a primary *Integrated Services Digital Network* (ISDN) connection. The central office switch (Class 5) is the entry point to the PSTN for any voice traffic, which does not terminate within the private PBX/N-ISDN network.

2.3 Virtual Private Networks (VPNs)

PBXs can be connected across an ATM network to form a *Virtual Private Network* (VPN). The VPN solution looks similar to the Enterprise network except that a service provider is managing the network. VPNs provide a network that is perceived as being private, but is actually part of a larger shared multiservice infrastructure maintained by a service provider. The multiservice voice VPN retains the intrinsic features of a private network and provides interconnectivity between the different locations of an Enterprise.

2.3.1 Enterprise VPNs

The Enterprise VPN could be based on one of the following:

- *Permanent Virtual Connection* (PVC) or *Soft Permanent Virtual Connection* (SPVC)[1] using either CAS or CCS
- *Switched Virtual Connection* (SVC) using CCS

2.3.2 PVC-Based Enterprise VPNs

Voice connectivity can also be provided through a frame relay access network or a cell relay access network connected to the ATM backbone.

ATM connections are established between a channelized frame relay linecard and an ATM linecard providing voice services. Service interworking takes place at the channelized frame relay card.

[1] An SPVC is a Soft Permanent Virtual Connection. The SPVC end points are usually determined by network management and the route of the SPVC is then set up by signaling.

2.3.3 SVC-Based Enterprise VPNs

VPNs can be SVC-based using cell relay access links and CCS (Q.931 or QSIG) for the *Time-Division Multiplexing* (TDM) trunks:

- Bandwidth usage is dynamic—that is, an ATM connection only exists while a call is in progress over the connection.
- Connections can be made directly between each location.
- It is not necessary to tandem connections through the headquarters' site.

Figure 2-2 represents a model for voice services from a PBX being delivered over an ATM network. Figure 2-3 illustrates the context for use of the terms "ingress" and "egress." Ingress refers to a cell stream flowing toward the switch core. Egress refers to flowing away from the switch core.

Voice access to the ATM network is either

- From a PBX over a *Narrowband ISDN* (N-ISDN) primary-rate connection offering 23 bearer channels, each at 64 Kbps (T1), or 30 bearer channels each at 64 Kbps (E1)

or

- From a PBX over a non-ISDN connection, that is, a T1 or an E1 trunk group

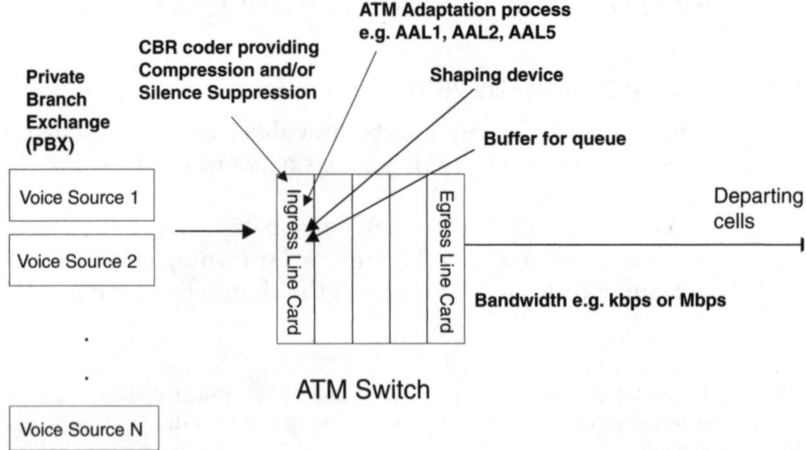

Figure 2-2 PBX to ATM voice traffic model

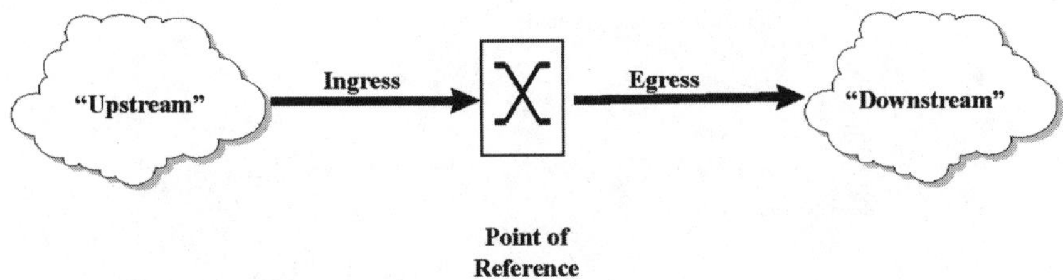

Figure 2-3 Ingress and egress

The voice model, as shown in Figure 2-2, assumes a number of channels encoded by *Pulse Code Modulation* (PCM) at 64 Kbps enter an ATM switch at a port of the ingress linecard. A *Constant Bit Rate* (CBR) coder at the ingress linecard receives the voice traffic from a PBX. The linecard, providing voiceband services, normally offers compression and silence suppression.

2.4 G.711 Coding

Today's telephone networks, whether public or private, operate synchronously. It was determined long ago that most speech energy lies somewhere between ~50 and 3,400 Hz, and that if the speech energy was band-limited to this range, good speech intelligibility would still be maintained.

Voice is encoded digitally by sampling the analog voice signal 8,000 times per second, once every 125 microseconds. This implementation is derived from Nyquist's theorem, which states that a signal of *Bandwidth* (B) must be sampled at a rate of 2B in order to capture all of its information. In PCM, each sample is quantized to one of 256 signal levels, centered about zero. The analog voice stream is encoded and digitized according to μ-Law or A-law, as described in ITU-T Recommendation G.711.[2] This generates a 64-Kbps data stream, whether speech is present or not.

The usual practice in a public network is for 30 64-Kbps PCM signals to be combined and multiplexed on a single 2.048-Mbps digital signal called an E1 (see Figure 2-4).[3] In the United States,

---

[2] If μ-law is used on one side of the ATM network and A-law is used on the other side, connections can be established where the format of the PCM signals will be converted automatically. For example, it would be possible to interconnect E1 and T1 networks where the E1 side is using A-law and the T1 side is using μ-law.

[3] The other two timeslots are used for framing and signaling—these are not PCM encoded.

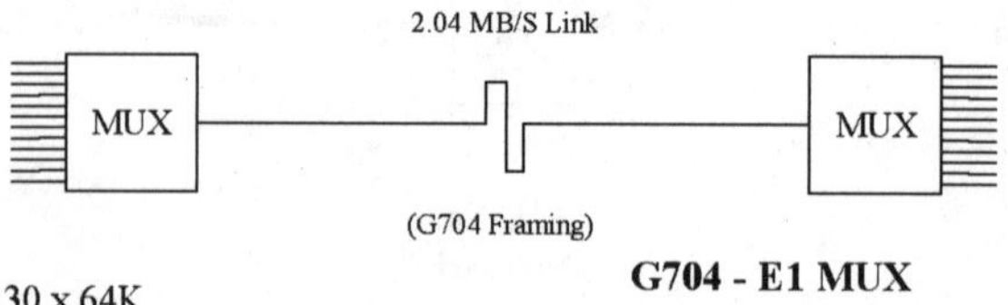

Figure 2-4 G704: 30 channels at 64 Kbps are multiplexed into an E1 signal at 2.048 Mbps.

the figure is 1.544 Mbps, because there are 24 circuits to a digital signal which is called a T1. Within a private network, you can use any framing method, but in the public networks, E1 frames follow a standard CCITT recommendation, G.704[6]. This is a byte-interleaved frame.

Overheads will be incurred through framing and cellification of the voice traffic. PBX signaling may be interworked at the edge of the ATM network or transported transparently.

2.5 Signaling Information

A voice call consists of two parts: the voice samples and the signaling information. The signaling information includes the dialed number, the on-hook/off-hook status of the call, and possibly a variety of other routing and control information. This signaling can be encoded in a number of ways and may be sent as CCS, CAS, or *Dual Tone Multiple Frequency* (DTMF) dialed digits.

Typically, multiple voice channels are combined into a single circuit. So for example, a European 2-Mbps E1 circuit contains 30 discrete voice channels, or an American 1.5-Mbps T1 circuit contains 24 channels. Signaling information may be embedded within each discrete voice channel (CAS) or aggregated into a single signaling channel, containing signaling information for all the channels on the circuit (CCS).

2.5.1 Channel Associated Signaling (CAS)

In CAS, each T1 channel has its own private signaling subchannel arranged by allocating the least significant bit of every sixth byte for signaling purposes, so five out of six samples are eight bits wide and the other one is only seven bits wide.

In CAS on an E1 link, four bits of signaling (ABCD) per time slot are multiplexed into a fixed portion of timeslot 16 as defined in G.732. This mode is also known as PCM 30.

CAS involves periodic transmission of AB- or ABCD-signaling bit patterns rather than signaling messages. Even though the patterns are transmitted in a common timeslot for an E1, it is called CAS rather than CCS because it is not message based. CAS is carried transparently across an ATM network within the same ATM *Virtual Channel Connection* (VCC) as the voice channel.

2.5.2 QSIG and DPNSS

QSIG and DPNSS are the main signaling protocols used between PBXs. QSIG is a private ISDN peer-to-peer protocol. The standards for QSIG were developed by the *European Computer Manufacturers Association* (ECMA). ITU-T Q.93x series recommendations provide for the basic services and generic functions. ITU-T Q.95x series recommendations provide for the supplementary services.

QSIG was designed to supersede the proprietary DPNSS, and QSIG is now the most prevalent. (See Appendix E and G for more detail on DPNSS and QSIG.) Both QSIG and DPNSS are examples of CCS.

2.5.3 Common Channel Signaling (CCS)

CCS is a message-oriented signaling protocol. Signaling messages for channels from one or more T1/E1 ports are communicated over a single common signaling channel. Thus, CCS is a method of signaling where the signaling channel is separate from the voice, and the signaling channel is carried independently of the voice.

2.5.4 Private Signaling System No. 1 (PSS1)

The standards for QSIG have been adopted internationally by the *International Standards Organization* (ISO), and this international private signaling system is called *Private Signaling System No. 1* (PSS1).

2.5.5 Narrowband ISDN and Q.931/DSS1

A network of PBXs can be grouped together to form a private N-ISDN using QSIG. QSIG is based on, and is very similar to, Q.931.

The Q.931 protocol enables digital access to the PSTN for all calls—local, national, or international—which do not terminate within the private PBX/N-ISDN network.

Q.931 (03/93) describes the *Digital Subscriber Signaling System No.1* (DSS1)–ISDN *User Network Interface* (UNI) layer 3. Q.921 specifies the DSS1 UNI layer 2.

2.6 Transporting PBX Signaling

The bandwidth requirement of an uncompressed voice channel from a PBX is 64 Kbps. The addition of framing overheads incurred through cellification at the ATM interface will increase that bandwidth requirement. The size of the increase will depend on the type of adaptation layer in use and the associated methods for encapsulating voice packets within the adaptation layer frame. PBX signaling must also be transported in addition to the voice bearer channels. There are two entirely separate methods of transporting PBX signaling. PBX signaling may either be transported passively and transparently across the ATM network, or alternatively, it is possible to interwork PBX signaling with the ATM signaling at the boundary of the ATM network. If the ATM connections are PVCs or SPVCs, signaling is transported transparently. If the ATM connections are SVCs, signaling is interworked at the boundary of the ATM network.

2.6.1 Transporting Signaling Transparently

If CAS is used, the signaling is carried within timeslot (channel) 16 and will not incur the provision of a further VCC within the ATM network—that is, the signaling is carried in the same ATM VCC as the voice bearer traffic. If CCS is used, the PBX signaling is carried separately from the voice signal and will incur the provision of one VCC for each port[4] or bundled group of channels.

DPNSS[5] and QSIG[6] are both CCS type protocols. Thus, the most likely scenario is that extra bandwidth for signaling will have to be provisioned for, unless SVCs are used to transport the voice across the ATM network. However, currently, PVCs and SPVCs are more commonly used in an ATM network.

---

[4] This is usually one E1 port.

[5] DPNSS is covered in detail in Appendix F.

[6] QSIG is covered in detail in Appendix G.

When CAS is used, it travels in the same ATM VCC as the voice. When CCS is used, for example, QSIG, then an additional VCC to carry the CCS signaling will be needed for each of the ports carrying the voice channels. This VCC can be a PVC or an SPVC.

A reasonable model for research purposes would assume CCS signaling and PVC or SPVC connections. An additional bandwidth of approximately 21.2 Kbps (50 cells per second) per 30 channels will be needed for signaling.

2.6.2 Interworking Signaling

An ATM core can replace TDM-based tandem/transit switches (Class 4) interconnected by a network of TDM trunks. Traffic is carried over ATM SVCs. Figure 2-5 shows how tandem switch functionality is provided over a broadband platform for international long-distance and local dialing using local interworking of signaling and integrated call processing.

Another approach would be to use a call server connected to the ATM network. The call server dynamically establishes an SVC through the ATM network as soon as it receives an SS7 signaling message requesting a call setup from the TDM side of the network. This application can also provide a suite of enhanced service capabilities including toll free, 900 services, local number portability, calling-card capabilities, and other intelligent network services.

The ATM network call setup is controlled by a nonintegrated call server. The *Private Network Node Interface* (PNNI) routing protocol sets up a route dynamically across the ATM portion of the network.

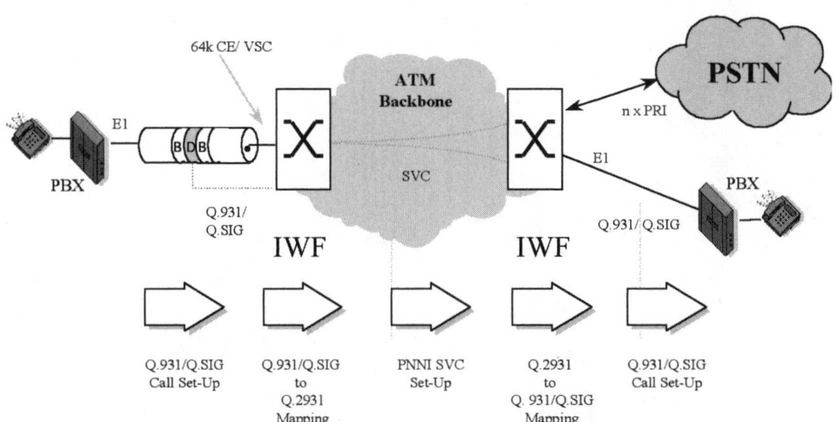

Figure 2-5 Signaling and routing in an ATM SVC network

Chapter

3

Mathematical Modeling of Voice Traffic

3.1 Statistical Gain in the ATM Core

Two aspects of telephony can be referred to as ON/OFF sources. For example,

- A voice channel is an ON/OFF source—that is, it is ON during a telephone call.
- Within any telephone conversation, there are alternating ON/OFF states of talk and silence.

There is a great potential for statistical gain in transmitting voice over ATM/IP networks through

- Idle channel detection, meaning that inactive channels are not sampled and transmitted
- Coding for compression, reducing the bandwidth required by a voice signal by up to a factor of eight times[1]
- Silence detection applied to an individual voice conversation to suppress the transmission across the network of packets carrying silence[1]

[1] The second and third bullet of this list may only be realized in the core of the network —for example, they will not apply to the first and last hop in the case where the voice data is returned to its uncompressed state complete with periods of silence.

- Transmission of the voice at a *Variable Bit Rate* (VBR)[2] class of service reducing the reservation of bandwidth closer to the average rate rather than continuously at the peak rate (as in circuit emulation)
- Multiplexing of several *Permanent Virtual Connections* (PVCs) into a *Permanent Virtual Path* (PVP) in the core
- The replacement of PVCs and PVPs with *Soft Permanent Virtual Connections/Switched Virtual Connections* (SPVCs/SVCs) and *Soft Permanent Virtual Paths/Switched Virtual Paths* (SPVPs/SVPs) for dynamic bandwidth usage rather than fixed bandwidth

The sizing of bandwidth for an ATM virtual circuit is performed in the switch by a Call Admission Control Algorithm.

This book describes a method for reducing the bandwidth due to silence removal to show how much more aggressive you can be when setting the traffic descriptors for silence suppressed voice prior to applying the Call Admission Control Algorithm. The analysis shows how many sources should be allowed with a fixed buffer size and a specified small cell loss probability.

The aim at this stage is the characterization of real-time conversational voice at the source. This section examines results from research and investigation into the following areas:

- The effects of *hangover* and *fill-in* in joining together very small silences—for example, pauses
- The characterization of the average talkspurt duration and average silence duration
- The arrival distribution of packetized voice
- The superposition arrival distribution for several voice sources
- The queue length distribution at the egress buffer of the ingress linecard

3.2 Measurement of Conversational Voice

Speech detectors operate with a hangover or a fill-in to bridge short gaps. The purpose of hangover and fill-in is to link together short silence periods in speech.

[2] VBR is an ATM class of service as defined by the ATM Forum.

Fill-in and talkspurt hangover are used in voice communication systems to achieve a reduction of signaling overhead and ensure more continuous flow of speech, but the use of fill-in and hangover can result in increased data delays. A hangover time of 150 to 250 *milliseconds* (ms) is usually used for conventional speech detectors. This significantly lengthens an average talkspurt duration.[3]

Gruber's paper [39] deals with the measurement and calculation of various speech temporal parameters of interest in an environment where speech activity detection is employed. The significance of Gruber's work in this paper lies in the important role that talkspurt hangover or fill-in play when characterizing voice.

Gruber shows that compared to hangover, fill-in results in longer-than-average silence and somewhat shorter-than-average talkspurt durations. Also, it is shown that fill-in results in the same talkspurt rate and somewhat lower speech activity. Apart from these points, you would not expect a great deal of difference in performance for speech interpolation of packet voice systems for either of these strategies.

3.2.1 Gruber's Results

Gruber's results are shown in Table 3.1. Gruber shows that a hangover of 150 to 200 ms[4] is all that is warranted to eliminate short silences. The talkspurt rate decreases by a factor of 10 from 200 per minute with no hangover to 20 per minute with 200 ms hangover for monologue.

Gruber states that it is unrealistic to assume that speech silence durations are purely geometrically or exponentially distributed. Some bunching of the hangovers occurs.

This was due to the lack of memory of the geometric probability density function for longer silence durations. Thus, for very long

TABLE 3-1 Gruber's Results [39]

	Hangover 202.5 ms	Fill-In 202.5 ms
Mean silence duration	606.3 ms	808.8 ms
Mean talkspurt duration	2.360 sec	2.157 sec

[3] The hangover time can be shortened to 32 ms by making the detector sensitive to the low-power speech signal above −51 dBm while keeping the high noise immunity [39].

[4] Note that in all cases ms refers to milliseconds.

hangovers, the truncation destroyed the lack of memory property. The sample does suggest that for the range of practical interest, it can be assumed that the probability density function for longer simulations will behave in a "memoryless" way.

3.2.2 Brady's Results

Brady [30] makes a statistical analysis of speech and shows that the level of the speech detector used causes results varying between oversensitivity to noise or insufficient speech detection. Brady used a system where a flip-flop was examined and cleared every 5 ms with the output being 1 if the speech detection threshold was crossed. The resulting string of 0's and 1's was examined. All talkspurts less than 15 ms were erased. All gaps less than or equal to 200 ms were filled into account for momentary interruptions. Thus, any two consecutive talkspurts that were separated by a silence duration of 200 ms or less were bridged together. Brady [31] recorded some statistical measures of eight conversations. Over a total conversation time of 51.56 min, the mean talkspurt length was found to be 1.34 sec with the mean silence duration found to be 1.67 sec.

Brady observes that the simplest Markovian model predicts that each state of talk and silence will have an exponential distribution where the mean activity factor of the talkspurt period is $1.34/3.01 = 0.445$ and the mean activity factor of the silence period is $1.67/3.01 = 0.555$.

3.2.3 Sriram's Results

Sriram et al. [52] utilize the results from Brady and Gruber in the analysis of two schemes for multiplexing voice and data.

Brady's results for the talkspurt and pause, which correspond to the use of the -35 dBm speech detector for the male voice, are equated to a mean talkspurt $\cong 1.2$ sec and a mean pause $\cong 1.8$ sec, where a value of less than or equal to 200-ms fill-in is applied.

Gruber's results for the talkspurt and pause corresponding to a voice in conversational speech with no fill-in are equated to a mean talkspurt of $\cong 170$ ms and a mean pause of $\cong 410$ ms.

Sriram et al. assume the following:

- The interactive data traffic follows a Poisson arrival pattern.
- The holding times of silence and pause are both exponentially distributed with a mean, as the one previously mentioned.

- The message length in terms of number of packets is assumed to have a general distribution.
- The frame duration is T = 10 ms, that is, where λ is the rate of the Poisson data message arrival process, and λT is the expected number of data packets received per frame.

Data delays were compared for the case where the conversations were in Japanese to the case where they were in English. It was observed that English speakers have longer talkspurt/silence durations as compared to Japanese speakers, and that for the same systems' configuration, the data delay is smaller for the case when the speech traffic is from Japanese talkers as compared to the case when it is from English speakers.

Sriram et al. suggest that because the data delays are smaller when the talkspurt/silence durations are smaller, the fill-in/hangover durations in speech should be kept as small as possible for efficient performance of a multiplexer with speech activity detection.

3.2.4 Heffes and Lucantoni's Results

Heffes and Lucantoni's [43] approach approximates the aggregate arrival process by a simple correlated nonrenewal stream, which is modulated in a Markov manner. The model chosen is the *Markov Modulated Poisson Process* (MMPP), a doubly stochastic Poisson process where the rate process is determined by the state of a continuous time Markov chain.

The packet stream from a single voice source is modeled by arrivals at fixed intervals of T ms during talkspurts and no arrivals during silences.

In particular, the packet arrival process from a single voice source is considered to be a renewal process with interarrival time distribution, F(t), given by

$$F(t) = [(1 - \alpha T) + \alpha T(1 - e^{-\beta(t-T)})]U(t - T)$$

where U(t) is the unit step function and the *Laplace Stieltjes Transform* (LST)

$$\bar{f}(s) = \int_{0}^{\infty} e^{-st}\, dF(t) = [1 - \alpha t + \alpha T + \alpha T\beta(s + \beta)^{-1}]e^{-sT} \quad \text{(3-1) [43]}$$

The mean packet arrival rate from a single source, λ, is clearly given by

$$\lambda = -1(\overline{f}'(0) = 1/(T + \alpha T/\beta)) \qquad (3\text{-}2)\ [43]$$

The parameters used in this paper are given by $\alpha^{-1} = 352$ ms, $\beta^{-1} = 650$ ms, and T = 16 ms.

Heffes and Lucantoni's approach corresponds to a geometrically distributed number of voice packets (with mean $1/\alpha$T) during an approximately exponentially distributed talkspurt with the mean duration α followed by an approximately exponentially distributed silent period with the mean duration β.

Data packets are assumed to arrive as a Poisson process. The results are compared to a Poisson curve. Results show that the *Standard Deviation* (SD) of delay for the simulation, Poisson and MMPP, are almost identical up to 100 voice lines. The Poisson results are much lower than the simulation and the MMPP method beyond 100 lines.

Example At 120 lines, Poisson estimates the SD of the delay at approximately 1 ms. The simulation and the MMPP method estimate the SD of delay at 2 to 3 ms. In further analyses, the average delay versus data utilization is plotted. The Poisson curve estimates the average delay at 10 ms for 40 percent utilization. The simulation and the MMPP method estimate the delay at 10 ms when 35 percent utilization is reached.

3.2.5 Sriram and Whitt's Results

Sriram and Whitt [54] analyze a model of a multiplexer for packetized voice and data. A major part of the analysis is devoted to characterizing the aggregate packet arrival process resulting from the superposition of separate voice streams (with focus on voice).

The aim of their paper is to develop an understanding of the aggregate voice packet arrival process and investigate simple approximations to determine the packet delays for both voice and data. It is assumed that successive talkspurts and silence durations from a single voice source form an alternating renewal process—that is, all these time intervals are independent with each talkspurt and silence duration being of random length.

The most important assumption is that the number of packets in a talkspurt is geometrically distributed on the positive integers. Note that this is consistent with measurements indicating that talkspurts are approximately exponentially distributed. See Bibliography [28] and [35].

The following assumptions are made in the model used by Sriram and Whitt:[5]

- The successive talkspurts and silence periods form an alternating renewal process—that is, all these time intervals are independent with each talkspurt and each silence duration being of random length.
- The exponential mean talkspurt duration is 352 ms (approximate inverse of 2.8-sec estimated talkspurt duration).
- The exponential mean silence duration is 650 ms (approximate inverse of 1.5-sec estimated silence duration).
- The talkspurts are approximately exponentially distributed.
- The silences are approximately exponentially distributed.
- The voice packetization period is 16 ms—that is, the packet interarrival time is 16 ms.
- The mean number of packets per talkspurt is 352/16 = 22 and the voice-line activity (the fraction of time each voice source is in talkspurt) is 0.351.
- The probability of a packet arrival is 21/22 = 0.9545.
- Packet length depends on the coding scheme being used—for example, for 32-Kbps *Adaptive Differential Pulse Code Modulation* (ADPCM) coding and a packet interarrival interval of 16 ms, the packet size is 64 bytes.
- The square coefficient of variation (variance divided by the square of the mean) of an interarrival time in this renewal process is 18.1, which reflects the highly bursty nature of a single voice source.

The exact theoretical interarrival distribution in the aggregate packet arrival process is compared to an exponential distribution with the same mean for $n = 20$ and $n = 100$.

Sriram and Whitt's results suggest that the superposition arrival process, for fairly large n, affects the queue because the

[5] Portions have been reprinted with permission from Kotikalapudi Sriram and Ward Whitt, "Characterizing Superposition Arrival Processes in Packet Multiplexers for Voice and Data," *IEEE Journal on Selected Areas in Communications* SAC-4, no. 6 (September 1986).

TABLE 3-2 Sriram and Whitt Results [54]

	$n = 20$	$n = 20$	$n = 100$	$n = 100$
Time (ms) T	Stationary Interval $1 - F(t)\,n$	Exponential $-n\lambda te$	Stationary Interval $1 - F(t)\,n$	Exponential $-n\lambda te$
0.1	0.9591	0.9570	0.8044	0.8028
0.2	0.9198	0.9159	0.6468	0.6446
0.4	0.8457	0.8389	0.4175	0.4155
1.0	0.6558	0.6446	0.1110	0.1113
2.0	0.4260	0.4155	0.1117	0.0124
4.0	0.1743	0.1726	0.0001	0.00015
16	0.00027	0.00088	2.5×10^{-19}	5.5×10^{-16}

long-term covariances among interarrival times (that is, bunching) begin to affect the queue under higher traffic intensities, and the aggregate packet arrival process eventually becomes much more variable than a Poisson process.

In a further simulation of 100 voice lines, a total number of nearly 20 million packets arriving were considered. Sriram and Whitt's results show that the packet loss in the system should be comparable to that of an M/D/1/K model, where *K* represents the buffer size.[6]

Simulations were performed with different buffer sizes, and the proportions of packets lost were compared with the proportions lost when there is a Poisson arrival process. For buffer sizes of 8 and 10 packets, the Poisson blocking probability nearly coincides with what would be the case with a Poisson arrival process.

In fact, when the buffer size (*K*) is small, the packet loss may be even smaller than that of the M/D/1/K model—that is, a single interarrival time distribution may be even less variable than an exponential distribution.

For a buffer size of 61 packets, the actual proportion of packets lost is much greater than would be the case with a Poisson arrival process. Unfortunately, Sriram and Whitt were unable to determine when the Poisson approximation ceases to be accurate.

They show that dependence among interarrival times can play an essential role. Their analysis of the multiplexer model shows that the aggregate packet arrival process possesses exceptional long-term positive dependence, partially characterized by the

[6] Further explanation of queuing theory is provided in Section 7.3.

indexes of dispersion, and that this dependence is a major cause of congestion in the multiplexer queue under heavy loads. Even though the aggregate packet arrival process is nearly Poisson, it is not appropriate to simply model the aggregate packet arrival process as a Poisson process with the correct rate. An approximation for the arrival process depends on the time scale, and the relevant time scale, in turn, depends on the traffic intensity in the queue.

For future research, Sriram and Whitt recommend the development of approximations for congestion measures in a queue when the arrival process is partially characterized by its average arrival rate and the index of dispersion should become a standard measurement tool in performance analysis.

Conclusions of Sriram and Whitt show that

- The interarrival time of packets is nearly exponential.
- The superposition of n independent and identically distributed renewal processes each with rate λ/n tends to a Poisson process with rate λ as n tends to infinity.

This statement is qualified with the proviso that a key condition is that the processes being superimposed become increasingly sparse as n increases.

- A Poisson approximation does work well under light to moderate loads.
- A Poisson approximation for the arrival process seriously underestimates delays under higher loads where the long-term covariances matter.
- The extent to which a superposition process is nearly Poisson depends not only on n, but also on the relevant time scale.
- Over short intervals of time, the superposition process looks like a Poisson process; in fact, over short intervals, the superposition process looks like something less variable than a Poisson process.
- Over longer intervals of time, the superposition process significantly deviates from a Poisson process and is highly variable.

3.2.6 Cheng's Results

Cheng et al. [25] confirm Sriram and Whitt's [54] observations that long-term correlation properties cannot be neglected, even if the number of sources is to be made large, and they further progress the work of Heffes and Lucantoni [43] to find matching parameters from real traffic. Their fundamental idea is based on the axiom that the moments of the interarrival time measured from the arrival traffic will completely describe the statistical characteristics of the arrival pattern.

A moment-generating function is proposed by Cheng et al. [25] where the mean delay of the MMBP (Poisson)/Geometric/1 queue for the original arrival process and that of the reconstructed process are both evaluated by queuing analysis. In six experiments, the main service rates are assumed to be 0.8 to give different utilization levels (defined as *mean arrival rate / service rate*). Cheng et al. then tested their method, the MMBP with bulk arrival.

The nth moment of a random variable, X, is defined as $M_n \equiv$

$$\lim_{k \to \infty} \frac{\sum_{i=1}^{k} (X_i)^n}{k} \qquad \text{3-3 [25]}$$

where X_i is the ith sample from an independent measurement.

Expressions for the first four moments are as follows:

$$M_1 = E[X]^{(1)} = \pi(1)$$

$$M_2 = E[X]^{(2)} = \pi(1)^{(2)} + \pi(1)^{(1)}$$

$$M_3 = E[X]^{(3)} = \pi(1)^{(3)} + 3\pi(1)^{(2)} + \pi(1)^{(1)}$$

$$M_4 = E[X]^{(4)} = \pi(1)^{(4)} + 6\pi(1)^{(3)} + 7\pi(1)^{(2)} + \pi^{(1)} \qquad \text{3-4 [25]}$$

The experiments show that knowing the values of the first four moments captures the key characteristics of the original arrival process faithfully, as can be verified by comparing the mean delays experienced by cells in both arrival processes.

This approach can generally be applied to matching other types of arriving processes, but the success of this method depends on the processes involved, and more moments may be required to yield a good match.

Conclusions

- An exponential distribution of speech and silence holding times is a reasonable model with the proviso that hangover (or fill-in) will affect the behavior of the distribution.
- Brady [31] shows that an average talkspurt in the English language lasts 1.34 sec, and an average silence duration lasts 1.67 sec. The inverse of each of these amounts will each form the mean of an exponentially distributed duration. Brady's method of analyzing voice corresponds very well to the behavior of the Constant Bit Rate (CBR) coder modeled in this book. Brady effectively resolves speech into 5-ms segments, and this is the basis of a single unit of output of a typical coder (for example, G.726 uses 5 ms and G.729 uses 10 ms).
- The silence removal software applied by the coders referenced in this book uses a hangover of approximately 200 ms. This appears to match the hangover size used by Gruber [39] and Brady [31]. (Note: Gruber is dealing with monologue speech rather than conversational speech, which is the subject of this book.)

  Gruber shows that the size of the 200-ms hangover being used here should not affect the "memoryless" property of the exponential arrival and duration of talkspurts and silences. However, beyond 225 ms, some bunching of hangovers may occur, and this would mean that an absolutely pure exponential, or geometric, speech or silence duration could not be assumed.
- Sriram and Whitt's [54] results indicate that the aggregate packet interarrival process can be modeled reasonably well by the exponential distribution for a low to medium load, and this means between 20 and 100 sources. When the number of sources is increased, the arrival process becomes more variable than Poisson. When ATM Adaptation Layer 2 (AAL2) is used, as many as 248 channels may contribute to the filling of the same adaptation layer frame, which will become a single ATM Virtual Channel Connection (VCC). In ATM Adaptation Layer 5 (AAL5), every single voice channel is normally represented by a separate VCC sized and queued individually, but any number of VCCs can be combined in a Virtual Path (VP). Thus, Sriram and Whitt's observations show that when sizing connections comprising of more than 100 channels, the exponential arrival distribution is possibly no longer a valid model.

- Heffes and Lucantoni [43] estimate the number of packets (cells in this case) in a talkspurt using a geometric series. Heffes and Lucantoni [43] also show that the Poisson results are much lower than the simulation, and the MMPP method beyond 100 lines.
- Sriram and Whitt [54] highlight the fact that adjustment of the talk and silence durations is necessary for different languages and that these parameters also affect the delay incurred in the arrival process.
- Alternative approaches described by Cheng et al. [25] use methods of capturing key characteristics of arrival traffic to build an equivalent traffic model.

Chapter

4

Transporting Voice over ATM

4.1 Adaptation to ATM

At the ingress point to the *Asynchronous Transfer Mode* (ATM) core network, the voice signal is encapsulated into an *ATM Adaptation Layer* (AAL) frame that is segmented into small, evenly sized packets called *cells*. At the egress buffer of the ingress linecard, the cells are queued and transmitted out toward the egress linecard, which possibly offers a higher-speed line, and then transmitted into the network.

The speed of the line and the size of the egress buffer at the ingress linecard determine the speed at which the queue is emptied and the amount of delay incurred waiting in the queue. The speed of the outgoing line at the egress port may typically vary from a T1 connection at 1.544 Mbps to an E1 at 2 Mbps, an E3 at 34 Mbps, a DS3 at 45 Mbps, or an STM-1 at 155 Mbps. Even higher rates such as an STM-4 at 622 Mbps are becoming common.

4.2 Alternative Adaptation Layers for Voice over ATM

The AAL supports all of the functions required to map information between the ATM network and the non-ATM application that may be using it.

Currently, three different adaptation layers may carry voice services over the ATM backbone. (AALs are standardized in the ITU-T I.363.x Series of Recommendations [14] [15] [18].)

This section explains the framing structure of each adaptation layer in order to make clear which adaptation layers will permit statistical gain and the overheads associated with the choice of an adaptation layer:

- **AAL1 (per I.363.1)**[14] Supports synchronous transmission, for example, circuit emulation for voice
- **AAL2 (per I.363.2)**[15] Designed for short packets such as in mobile voice scenarios
- **AAL5 (per I.363.5)**[18] Can be interworked with *Frame Relay* (FR)

Both AAL2 and AAL5 allow for the application of voice compression and silence suppression to reduce the bandwidth required to transport a voice channel. AAL1 is designed to transmit voice channels at a constant, unreduced rate.

4.3 Circuit Emulation Using AAL1

AAL1 has been standardized in both the ITU-T and ANSI since 1993, and is incorporated in the ATM Forum specifications for *Circuit Emulation Services* (CES), structured and unstructured. ITU-T I.363.1 [14] provides the specification for *ATM Adaptation Layer 1* (AAL1).

Circuit emulation enables a traditional point-to-point connection, such as a T1 or an E1 circuit between *Time-Division Multiplexer* (TDM) nodes, to be realized over an ATM backbone. This is accomplished by the adaptation of the digital bit stream into a *Constant Bit Rate* (CBR) cell stream for transmission through the ATM network, followed by reassembly of the cell stream into the original data for transmission out of the network to a terminating device.

4.3.1 AAL1 Packet Data Unit (PDU) Format

The format of the AAL1 PDU is shown in Figure 4-1.[1] The convergence sublayer in the upper half of the AAL1 uses 1 byte of the 48-byte payload of the ATM cell for control information, therefore

[1] Reproduced with the kind permission of the International Telecommunication Union. The responsibility for selecting extracts lies with the author of this book and in no way can be attributed to the ITU.

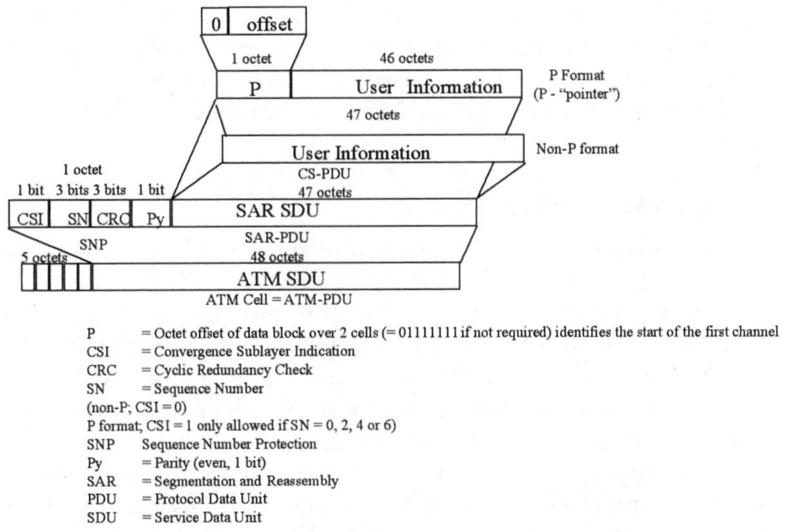

Figure 4-1 AAL1 PDU format[3]

leaving the cell payload to be filled with 47 bytes or 376 bits of data. Of these 47 bytes, another byte is used for a pointer once in every eight cells for all cases of structured mode, except when a single channel without *Channel Associated Signaling* (CAS) is being transported. Each cell has an overhead of 6 bytes in total, which is 5 bytes of ATM cell overhead plus 1 byte of overhead from the AAL1.

As far as the end nodes in the circuit are concerned, the fact that the intermediate physical circuit is passed through an ATM network is transparent. Circuit emulation over AAL1 offers a CBR class of service—that is, a constant amount of bandwidth is reserved across the ATM network.

4.3.2 Unstructured Circuit Emulation

Unstructured circuit emulation is the ATM-Forum-specified[2] approach based on using an AAL1 CBR ATM connection to carry

---

[2] The ATM Forum is a nonprofit making international organization accelerating industry cooperation on ATM technology. The ATM Forum does not, expressly or otherwise, endorse or promote any specific products or service.

[3] PDU format is used in all cases of structured mode except when a single channel without CAS (Basic Mode) is being transported.

a full T1 or E1 circuit, or full T3 and E3 circuits, between two points in the network. This connection may typically be used to carry end-to-end circuits between TDMs or digital *Private Branch Exchanges* (PBXs) over the ATM backbone transparently.

4.3.3 Structured Circuit Emulation

Structured circuit emulation establishes an AAL1 $N \times 64$ Kbps circuit, such as a fractional T1 or E1, over the ATM backbone. Individual groups, or bundles, of timeslots are mapped to ATM circuits. Because the $N \times 64$ Kbps service can be configured to use a fraction of the timeslots or channels available on the service interface, it is possible to allow several independent emulated circuits to share one service interface.

The capability of allowing several AAL1 entities to share one service interface, where each AAL1 entity is associated with a different *Virtual Channel Connection* (VCC), enables functional emulation of a DS1/DS0 or an E1/DS0 digital crossconnect switch. In structured mode, the CAS associated with each of the channels in the $N \times 64$ Kbps bundle is also transported within the AAL1 structure and over the same ATM connection that is transporting the bundle.

4.4 Bandwidth Requirements for Circuit Emulation

The bandwidth (cells per second) required to carry the voice traffic depends on whether the CES being used is structured or unstructured.

4.4.1 Circuit Emulation in Unstructured Mode

In unstructured circuit emulation, all channels are collected and the cell fill rate cannot be adjusted. The following cell rates[4] for unstructured circuit emulation using AAL1 always apply:

[4] The T1/E1 cell rates for circuit emulation include 1 OAM cell of overhead (this is a performance-monitoring cell) per 128 user cells, plus 1 cell per second is also added for the *Alarm Indication System* (AIS). This footnote also applies to (4-2) and (4-4).

$$\text{E1 cell rate} = \left\{ \left(\frac{2.048\ (10^6)\ \text{bits/second}}{(47\ \text{octets/cell})(8\ \text{bits/octet})} (1 + 1/128) + 1 \right\} \right. \tag{4-1}$$

$$= 5{,}491 \text{ cells per second}$$

$$\text{T1 cell rate} = \left\{ \left(\frac{1.544\ (10^6)\ \text{bits/second}}{(47\ \text{octets/cell})(8\ \text{bits/octet})} (1 + 1/128) + 1 \right\} \right.$$

$$= 4{,}140 \text{ cells per second}$$

This is equivalent to an E1 bandwidth of 2,328.18 Kbps (average 75.1 Kbps per channel for 31 channels) and a T1 bandwidth of 1,755.36 Kbps (average 73.14 Kbps per channel for 24 channels).

4.4.2 Circuit Emulation in Structured Mode

For structured AAL1 data, the cell rate will depend on three factors:

- The configured cell fill level
- The number of channels being carried, which provides the cell rate
- The use of CAS

Each E1 connection is made up of 32×64 Kbps timeslots. (For a T1 connection, there are 24×64 Kbps timeslots.) Timeslot 0 is reserved for framing and alarm information. If CAS is used, time slot 16 is used to transport signaling, and 30 channels are available for data. Otherwise, if CAS is not being used, 31 channels are available, and this is generally known as *31-channel mode*, or as a BASIC service. Timeslots for transmission are referred to as *numbered channels*. An example of an E1 channel numbering scheme is shown in Figure 4-2.

Combining Timeslots in Structured Mode In structured circuit emulation, data channels from several 64-Kbps timeslots can be grouped together into a single ATM connection using one virtual connection. The number of timeslots collected together governs the amount of data that will be put into the AAL1 frame. Also, where more than 64 Kbps of bandwidth is required, this enables the aggregation of bandwidth from several timeslots.

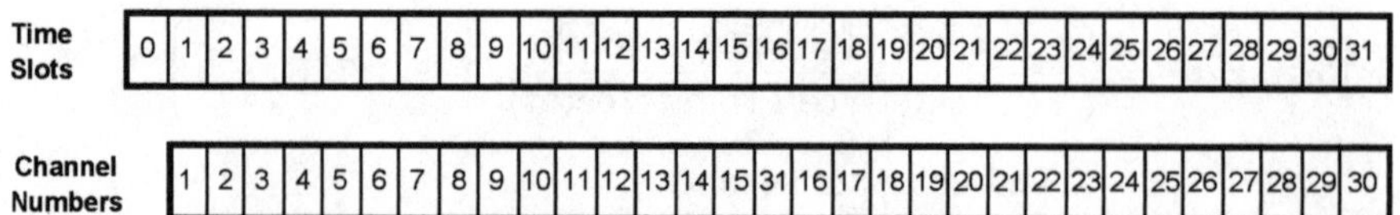

Figure 4-2 Timeslot to channel number mapping

Bandwidth Required for Structured T1 and E1 Ports The equation for calculating the cell rate for structured circuit emulation is the same for E1 and T1 ports. The cell rate for structured T1 and E1 basic service is

$$\text{T1 and E1 BASIC cell rate} = \left\{\left(\frac{8000N}{K}(1 + 1/128)\right) + 1\right\} \text{cells/sec} \qquad (4\text{-}2)^5$$

Table 4-1 provides a spreadsheet to calculate the cell rate for structured circuit emulation using Equation 4-2. The formula for Table 4-2 is included in Appendix H.

Less delay will be incurred if cells are only partially filled. When CAS is used, the cell rate is higher due to the addition of the CAS bits at the end of every multiframe in the SAR-SDU.

The equations for calculating the cell rate are as follows:

$$\text{T1 CAS cell rate} = \left\{\begin{matrix} 8000\left\lceil\frac{49N}{48K}\right\rceil, \textit{for N even} \\ 8000\left\lceil\frac{49N+1}{48K}\right\rceil, \textit{for N odd}\end{matrix}\right\} \text{cells/sec} \qquad (4\text{-}3)$$

$$\text{E1 CAS cell rate} = \left\{\begin{matrix} \left\{\left(8000\left\lceil\frac{33N}{32K}\right\rceil(1 + 1/128)\right) + 1\right\} \text{cells/sec, } N \textit{ even} \\ \left\{8000\left\lceil\frac{33N+1}{32K}\right\rceil(1 + 1/128) + 1\right\} \text{cells/sec, } N \textit{ odd}\end{matrix}\right\} \qquad (4\text{-}4)^6$$

[5]K = The number of bytes (octets) included in each cell's payload, N = The number of channels

[6]K = The number of bytes (octets) included in each cell's payload, N = The number of channels

TABLE 4-1 Cell Rate Calculator for Structured Circuit Emulation

The Number of Channels (1–31)	Fill Level (Kb)	Cell Rate (Cells/sec)	Bandwidth (Kbps)
31	47	5,319	2,255.25

TABLE 4-2 Cell Rate Calculator for CAS CE

T1 and E1 CAS Service						
Connection Type	The Number of Channels (N)	Fill Rate (Kb)	Cell Rate N Even	Bandwidth (Kbps)	Cell Rate N Odd	Bandwidth (Kbps)
T1	24	47	4,204	1,782.496	4,208	1,784.19
E1	30	47	5,309	2,251.016	5,314	2,253.13

Table 4-2 provides a spreadsheet to calculate the cell rate for CAS circuit emulation using Equations 4-3 and 4-4. The formula for Table 4-2 is included in Appendix H.

Figures 4-3 and 4-4 illustrate the behavior of the cell rate for a T1 and E1 CAS service as the number of channels is increased.

4.5 Voiceband Services Using AAL2

ATM Adaptation Layer Type 2 (AAL2) had its beginnings in a contribution to Committee T1S1.5 entitled *Short Multiplexed AAL* (SMAAL) in September 1995, which was authored by John Baldwin of Lucent Technologies Ltd. SMAAL was introduced to the ITU-T at the May 1996 meeting of Study Group 13 in Geneva. AAL2 is defined in the ITU-T Recommendation I.363.2 [15] that was determined at the Study Group 13 meeting in Seoul, Korea, in February 1997 and was approved at the September 1997 Study Group 13 meeting in Toronto.

AAL2 is divided into two sublayers: the *Service Specific Convergence Sublayer* (SSCS) and the *Common Part Sublayer* (CPS). In ITU-T Recommendation I.363.2, the SSCS is defined as the link between the AAL2 CPS and the higher-layer applications of the individual AAL2 users. Several SSCS definitions are planned, but the SSCS is currently null.

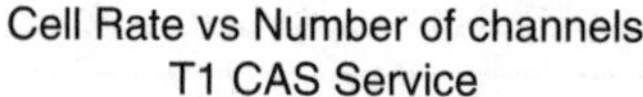

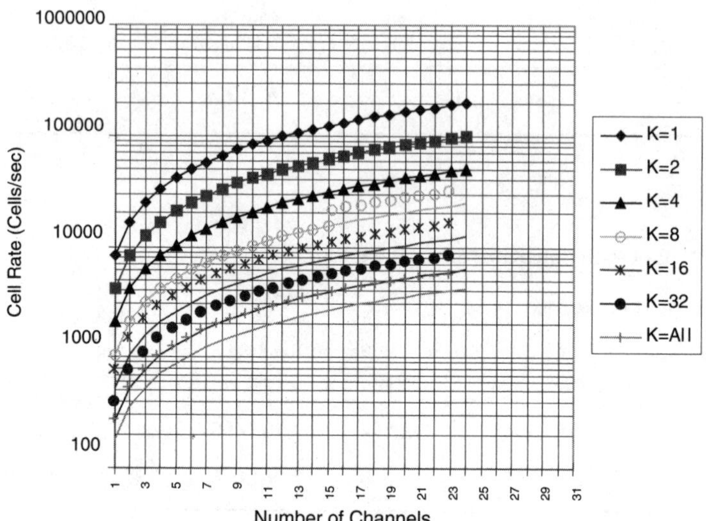

Figure 4-3 The cell rate versus the number of channels for T1 CAS service

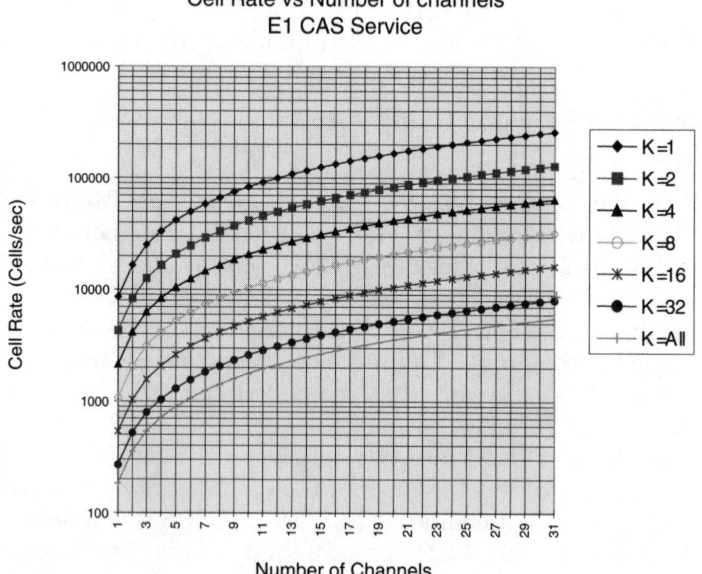

Figure 4-4 The cell rate versus the number of channels for E1 CAS service

4.5.1 AAL2 Common Part Sublayer (CPS)

The AAL2 CPS, as defined in I.363.2, provides the basic structure for identifying the users of the AAL, assembling/disassembling the variable payload associated with each individual user, error correction, and the relationship with the SSCS above it. The structure of AAL2 is shown in Figure 4-5.[7] AAL2's CPS possesses the following characteristics:

- It is defined on an end-to-end basis as a concatenation of AAL2 channels.
- Each AAL2 channel is a bidirectional virtual channel with the same channel identifier value used for both directions.
- AAL2 channels are established over an ATM layer *Permanent Virtual Connection* (PVC), *Soft Permanent Virtual Connection* (SPVC), or *Switched Virtual Connection* (SVC).
- The multiplexing function in the CPS merges several streams of CPS packets onto a single ATM connection.

The CPS PDU consists of a 3-byte CPS packet header and a payload, which defaults to 45 bytes, but can be extended to 64 bytes. The four fields of the CPS packet header are the *Channel Identifier* (CID), the *Length Indicator* (LI), the *User-to-User Identifier* (UUI), and the *Header Error Check* (HEC) field.[8]

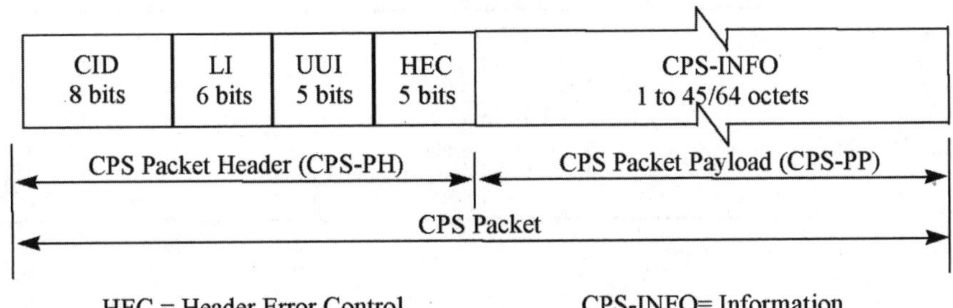

Figure 4-5 Structure of the AAL2 header

[7] Reproduced with the kind permission of the International Telecommunication Union. The responsibility for selecting extracts lies with the author of this book and can in no way be attributed to the ITU.

[8] The HEC checks the completeness and accuracy of the transmitted data.

- **CID field** As shown in Table 4-3, this uniquely identifies the individual user channels within the AAL2 and allows up to 248 individual users (values 8–255) within each AAL2 structure.
- **LI field** Identifies the length of the packet payload associated with each individual user and ensures conveyance of the variable payload. The value of the LI is one less than the packet payload, and has a default value of 45 octets or may be set to 64 octets.
- **UUI field** Provides a link between the CPS and an appropriate SSCS that satisfies the higher-layer application. Different SSCS protocols may be defined to support specific AAL2 user services or groups of services. The SSCS may also be null.

Coding of the CID field is shown in Table 4-3.

After assembly, the individual CPS packets are combined into a CPS-PDU payload.

Coding of the UUI field is shown in Table 4-4.

The Offset field, which is a subset of the Start field in the CPS-PDU, identifies the location of the start of the next CPS packet within the CPS-PDU. The Start field is protected from errors by the parity bit, and data integrity is protected by the sequence number.

TABLE 4-3 The Structure of the CID Field

CID Value	Use of CID Field
0	Not Used
1	Reserved for layer management peer-to-peer procedures
2–7	Reserved
8–255	Available to identify user channels

TABLE 4-4 The Coding of the UUI Field

UUI Value	Use of UUI Field
0-27	Identification of SSCS entries
28,29	Reserved for future standardization
30,31	Reserved for layer management (OAM)[a]

[a]OAM = Operation and Maintenance

4.6 Voiceband Services Using AAL5

Voice and data can also be transported over an ATM network within the payload of an AAL5 PDU. Within the AAL5 packet, another protocol layer is needed to differentiate the different types of information that can be carried over the connection, for example, voice packets, silence packets, and fax packets.

4.6.1 Voice and Data Within an FRF.11 Subframe

The *Frame Relay Forum's* (FRF) Voice over Frame Relay Implementation Agreement, FRF.11 [21], provides a suitable and standardized protocol for doing this. A direct translation is possible between FRF.11 within AAL5, and FRF.11 within an FR frame. This translation is defined by the transparent mode of the Frame Relay Forum's FR/ATM PVC Service Interworking Implementation Agreement, FRF.8 [20]. (FRF.11, FRF.8, and FRF.5 are explained in further detail in Appendix A.) An FRF.11 subframe, as shown in Figure 4-6, becomes the payload of an AAL5 PDU. Each subframe consists of a variable-length header and a payload. The subframe header identifies the voice/data subchannel, and if required, the payload type and length. An extension octet containing the most significant bits of the voice/data channel identification and a payload type is present when the *Extension Indication* (EI) is set. A payload-length octet is present when the LI is set. The payload type (optional) distinguishes between different types of information such as voice, silence, CAS, *Dual Tone Multi-Frequency* (DTMF),[9] and fax.

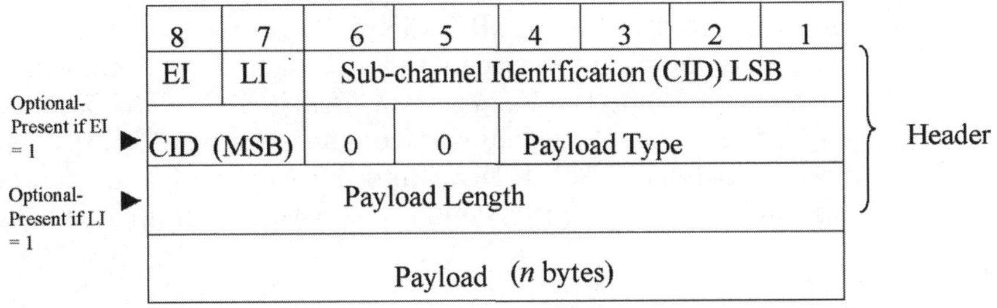

Figure 4-6 FRF.11 subframe format

[9] DTMF is a method of signaling using the speech transmission path. This method employs 16 distinct signals, each composed of two voiceband frequencies. These 16 tone pairs may also be called *touch tone signals*. DTMF is the usual form of address signaling used by *Plain Old Telephone System* (POTS).

The payload length does not need to be sent if this subframe is the only subframe or the last subframe in the FR frame or AAL5 packet.

The minimal subframe header is a single octet containing the least significant bits of the voice/data channel identification along with extension and length indications.

NOTE: *If the subchannel identifier is greater than 63 (the largest number that can be coded in the 6 bits available), the subchannel identifier will have its most significant bits carried in the second octet of the subframe header. This will result in an extra octet of overhead if the second octet was not already present.*

Example of FRF.11 Encoding in AAL5 If 2 bytes of overhead are assumed for the FRF.11 header, 1 byte containing the sequence number, and 8 bytes for the AAL5 trailer, a total of 11 bytes of overhead are incurred.

FRF.8 (FR/ATM PVC Service Interworking Implementation Agreement) [20] When the FR frame is mapped into an AAL5 PDU in the frame-to-ATM direction, the FR frame's inserted zero bits and CRC-16 are stripped.[10]

The Q.922 *Data Link* (DL) core frame header is removed, and some of the fields of the header are mapped into the ATM cell header fields. AAL5 provides message (frame) delineation and 32-bit CRC bit error detection. In the direction from ATM to FR, the message delineation provided by AAL5 is used to identify frame boundaries and zero bits; a CRC-16 and flags are added.

One-to-one mapping between the FR *Data Link Connection Identifier* (DLCI) and the *Virtual Path Identifier / Virtual Channel Identifier* (VPI/VCI) is used for service interworking (FRF.8).[11] The association between the FR DLCI and the ATM VPI/VCI is made at the time the PVC is provisioned. The association may be arbitrary or systematic.

---

[10] *Cyclic Redundancy Check* (CRC) is a field of 16 bits used to check the integrity of the data after transmission.

[11] See Appendix A for further details on FRF.11, FRF.8, and FRF.5.

Conclusions

- AAL1 provides a maximum and Constant Bit Rate (CBR), which does not allow for statistical gain in multiplexed voice channels and incurs a framing overhead of 1 byte.
- If a single uncompressed 64-Kbps channel is transported across an ATM network using AAL1, the bandwidth is increased to 75.1 Kbps (E1 unstructured and E1 CAS) or 72.76 Kbps (E1 Basic). This is an increase of 17.3 percent and 13.7 percent, respectively.
- AAL2 enables coder units from up to 248 channels to be combined into a single AAL frame. AAL2 incurs a minimum overhead of 3 bytes for each coded sample encapsulated in the AAL2 frame.
- AAL5 enables interworking with FR, which is a significant advantage as large networks tend to bring in low-speed data, such as voice, using FR in the access tier. Furthermore, AAL5 is currently the most used type of adaptation layer.
- AAL5 incurs an overhead of 8 bytes for the AAL5 trailer. FRF.11 incurs an overhead of 2 bytes. A further byte is used for a sequence number. A total of 11 bytes of overhead is incurred in each AAL5 frame when interworking with FR at the edge of the network. Within the AAL5 frame, there can be one or a number of coded units, depending on the packing factor selected, but all of these will derive from the same source channel.
- In all cases, an adaptation layer frame may be segmented into one or several ATM cells. Each ATM cell will incur a further overhead in the 5-byte ATM cell header.

Chapter

5

Voiceband Processing

Economic gains in voice transmission are made possible by two factors: low bit rate through compression and the exploitation of speech silence durations. Overestimation of the required bandwidth will cause underutilization of the network resources and increased call blocking probability. Underestimation can cause cell loss.

This section describes the bandwidth and delay characteristics of compressed and/or silence suppressed voice samples encapsulated within the framework of the *ATM Adaptation Layers* (AALs) AAL2 and AAL5. The structure of AAL1 does not allow for the application of compressed or silence-suppressed voice, and all voice signals are carried at a maximum and constant bandwidth. AAL2 and AAL5 (using FRF.11 and FRF.8) provide a structure that enables speech-processing functions to be applied to channels where the channel coding at the ingress DS1/E1 port is in the format of uncompressed 64-Kbps *Pulse Code Modulated* (PCM) (μ-law and A-law).

For example, Figure 5-1 illustrates a typical network scenario where *Asynchronous Transfer Mode* (ATM) linecards located at the T1/E1 interfaces of the ATM switches perform the following voiceband processing functions:

- Idle channel suppression is used for *Permanent Virtual Connection* (PVC) and *Switched Permanent Virtual Connection* (SPVC)[1] connections to suppress the sampling of timeslots from inactive voice channels.

---

[1] Idle channel suppression is not beneficial to *Switched Virtual Circuits* (SVCs), where a *Virtual Channel Connection* (VCC) is released when a channel is idle.

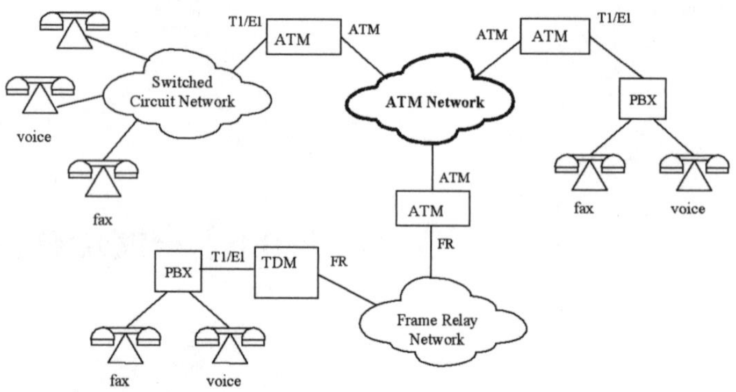

Figure 5-1 Typical voiceband services card application with voiceband processing

- Speech compression/decompression and silence removal are performed on channels carrying speech conversations to reduce the bandwidth consumed by these channels through the ATM network.
- Fax modulation/demodulation is performed on channels carrying fax calls to reduce the bandwidth consumed by these channels through the network.

5.1 Compression

Compression is a process where the received voice signal is reproduced using fewer bits. For example, 1.5 sec of PCM Clear Channel speech will be transmitted over a 64-Kbps channel using 96,000 bits. A G.729A [4] compression process can reproduce the same period of speech using 12,000 bits.

However, unlike normal data, which is simply sent as quickly as possible across the network, the timing of the voice signal cannot be substantially altered—for example, 1.5 sec of speech by the sender has to be delivered over 1.5 sec to the receiver.

Although compression reduces the amount of bandwidth needed to code the voice, it makes it more difficult to maintain this timing relationship because it has, in effect, slowed down the data rate going into the ATM network, which means that it takes longer for an ATM cell to be filled and transmitted.

To limit the effects of the *delay bottleneck* created by compression, most *Voice over ATM* (VoATM) implementations will sacrifice some of the bandwidth gain offered by compression and will restrict the amount of compressed data put into an AAL2/AAL5 frame before it is sent.

5.2 Clear Channel

When AAL1 is used, channels are always transported at the maximum possible bandwidth, that is, as Clear Channel. Even when AAL2 and AAL5 are used, compression may not be appropriate for some types of traffic, for example, where

- The data communications equipment requires a transparent data channel of the maximum possible channel bandwidth, as, for example, when *Dual Tone Multiple Frequency* (DTMF) tones are sent.
- The channels have already been converted into *Adaptive Differential Pulse Code Modulation* (ADPCM); or any other format, by upstream *Time-Division Multiplexing* (TDM) equipment.

In either case, channels are transported using a Clear Channel at the maximum possible bandwidth.

5.3 Compression Coder Units

Voice coders each produce units (quantums) of compressed data, which are referred to as *coder units*. The duration of compressed speech within a coder unit depends upon the specific coder and its underlying algorithm. Thus, the encoding of voice into an AAL frame takes place in units of *n milliseconds* (ms) depending on how many coder units are encapsulated in the frame. Table 5-1 shows the duration of a coder unit in milliseconds and the equivalent output in bits for Clear Channel and 9 different types of coder.

5.4 Combining Coded Units in an AAL Frame

The method of filling an AAL frame with coded voice units varies depending on the type of adaptation layer used, as follows:

- **AAL1** Structured circuit emulation uses AAL1, and this is always a *Constant Bit Rate* (CBR) service—that is, the entire PCM bandwidth is allocated for each contributing channel. This is 64 Kbps per channel, plus overheads added by framing and cellification. Single bytes from each of the contributing channels are interleaved onto a single ATM connection on a round-robin

TABLE 5-1 Duration (ms) and Size (bits) of 10 Different Types of Voice Coder Units

Coder	Reference Number of Standard if Applicable	Output Rate Coder (bits per second)	Coder Unit in n Milliseconds	Number of Bits in Coder Unit
Clear Channel	NA	64,000	5	320
G.711 μ-law	1	64,000	5	320
G.711 A-law	1	64,000	5	320
G.723.1 5.3 Kbps	2	5,300	30	159
G.723.1 6.3 Kbps	2	6,300	30	189
G.726 16 Kbps	3	16,000	5	80
G.726 24 Kbps	3	24,000	5	120
G.726 32 Kbps	3	32,000	5	160
G.726 40 Kbps	3	40,000	5	200
G.729A 8 Kbps	4	8,000	10	80

basis. Thus, the total bandwidth requirement of each channel is always increased above the 64 Kbps per channel received from the *Private Branch Exchange* (PBX) source. In unstructured circuit emulation, the contents of the whole trunk are loaded onto a single ATM connection. In Structured Circuit Emulation, the cell fill level may be adjusted to determine how many bytes of an ATM cell are filled with data before the cell is sent; thus, the cell fill level parameter controls the delay factor.

- **AAL2** When AAL2 is used, the coded units may be derived from several voice channels, and when more than one channel is involved, the coded units[2] are interleaved on a round-robin basis. The coded units from one or several channels are loaded into a single AAL2 frame, which can overlap more than one ATM cell. The combined use (Timer_CU) timer sets an upper bound on the number of coded units that may be inserted into the AAL2 frame. A single-byte Offset field at the beginning of an ATM cell indicates the beginning of the next AAL2 *Packet Data Unit* (PDU) within that cell. The bandwidth allocated to the received voice signals may be reduced when silence suppression and/or compression are applied.
- **AAL5** When AAL5 is used, the number of coder units included in one AAL5 frame depends on a configured packing factor. All

[2] In combining coded units onto the same AAL2 interface, the coded units may be derived from different types of coding processes.

of the coded units are derived from a single voice channel and are loaded sequentially into a single AAL5 frame that is segmented into an integer number of cells. Filler may be inserted into the last cell. The bandwidth allocated to the received voice channels may be reduced when silence suppression and/or compression are applied.

Table 5-2 summarizes these methods and the implications for bandwidth and delay of the different methods of filling the AAL frame.

5.4.1 Combining Coder Units into an AAL2 Frame

Example In AAL2, a coder unit may spread over more than one cell, which allows for greater efficiency in cell filling. A timer setting specifies the amount of time that may elapse before a cell is sent after the first coded unit has been inserted in it. The timer may allow for further coder units from any contributing voice channel to be inserted into the frame. Unfortunately, it is not possible to predict if coded units from different channels are going to arrive evenly spaced. Some coded units could be generated simultaneously from different channels. In the case where all the multiplexed channels are producing their coded units at exactly the same time, after the first coded unit is received that partly fills an AAL2 cell, there may be no more data received for the duration of a coder unit. Therefore, a worst-case setting for the timer is one coder unit—that is, within one more coder unit of time, the same channel, or one of the contributing channels, will have another coder unit ready for transmission. However, if the timer is set for too long increased delay could be imposed at the AAL2 multiplexing point if a sufficient quantity of coded voice is not available to fill a cell before the timer expires. This can happen when silence detection is in use and silence (i.e. no voice) is present on some of the channels that are being multiplexed together. Also the spacing of coded units being generated from a single channel, relative to other channels, might even shift over time due to silence detection activity (and depending on the internal implementation details of the voice coder).

Table 5-3 shows the average bandwidth requirements for AAL2 for a given number of channels and variable timer settings assuming that the coded unit arrival process is evenly spaced.[3] Note that

[3] Formulas for the derivation of the editable tables in this book are given in Appendix H.

TABLE 5-2 Comparison of Methods of Filling an AAL1, AAL2, and AAL5 Frame

Adaptation Layer Type	Bandwidth Allocated to Channel(s)	Method of Filling Adaptation Layer Frame	Implied Cell Delay
AAL1	**Unstructured:** **Constant**—Full E1 trunk @ 2.048 Mbps Full T1 trunk @ 1.544 Mbps **Structured:** **Variable**—Dependent on a multiple of 64 Kbps, plus overheads for each contributing channel. Structured Circuit Emulation has a configurable cell fill factor, which will affect the average bandwidth per channel.	Single bytes from one or several channels are inserted into the AAL1 frame on a round-robin basis.	**Unstructured:** Average cell fill delay 0.184 ms (E1) 0.244 ms (T1)[b] **Structured:** **Minimum**—Approximately 0.005 at very low cell fill rate 1 or 2 bytes **Maximum**—Approximately 6 ms for one channel at full cell fill rate
AAL2	**Variable**, depending on the following: ■ Type of coder ■ Number of channels contributing ■ Timer setting, which may or may not allow further coder units to be inserted before a cell is sent	■ Coder units from one or several channels are inserted into the AAL frame on a round-robin basis. ■ The timer setting determines the number of coded units that can be "waited for" before a cell is sent. ■ Filler may occur in any cell if a sufficient number of coder units cannot be gathered.	*Delay* depends on the number of coder units allowed by the timer setting divided by the number of channels. **Minimum**—One coder unit duration divided by the number of channels contributing. **Maximum**—45 bytes divided by the coder unit size of one channel multiplied by the coder unit duration.[a]
AAL5	**Variable**, but always a multiple of an integer number of cells, according to the following: ■ Type of coder ■ Length of coder unit in milliseconds ■ Packing factor ■ Cell fill factor	■ Coder units from a single channel are inserted into the AAL frame sequentially. ■ The packing factor determines how many units are "waited for" before a frame is segmented into an integer number of cells. ■ The cell fill factor determines the number of bytes put into the cell payload and thus the amount of filler in each cell.	**Minimum**—Duration of one coder unit (when the packing factor = 1) **Maximum**—Multiple of one coder unit duration multiplied by the packing factor

[a] The default length of the AAL2 frame is 45 bytes, but there is an alternative option of 64 bytes.

[b] In all cases, ms represents milliseconds.

TABLE 5-3 Bandwidth Requirements for AAL2

Coder 1	Output Rate of Coder (bps) 2	Coder Unit (ms) 3	Number of Channels Contributing to AAL2 Interface 4	Choose Max. Timer Setting (ms) 5	Recomm. Timer Setting for Full Cell Fill 6	Bandwidth Required for all Channels (Kbps) 7	Average Bandwidth per Single Channel (Kbps) 8	% Efficiency Based on Contributing Channels & Full Cell Fill 9
Clear Channel	64,000	5	1	5	5.00	77.58	77.58	100.00%
G.711 μ-law	64,000	5	10	0.4	0.50	848.00	84.80	91.49%
G.711 A-law	64,000	5	30	0.17	0.17	2,327.49	77.58	100.00%
G.723.1 5.3 Kbps	5,300	30	10	3	6.00	68.79	6.88	97.34%
G.723.1 6.3 Kbps	6,300	30	30	0.5	1.00	424.00	14.13	56.65%
G.726 16 Kbps	16,000	5	10	1	1.50	234.55	23.46	82.98%
G.726 24 Kbps	24,000	5	10	1	1.00	324.77	32.48	100.00%
G.726 32 Kbps	32,000	5	10	0.75	1.00	414.98	41.50	97.87%
G.726 48 Kbps	48,000	5	30	0.17	0.17	1,786.21	59.54	100.00%
G.729A 8 Kbps	8,000	10	10	5	3.00	117.28	11.73	100.00%

the setting for the timer that provides the best compromise between most cells being filled and minimizing delay will be somewhere in between the setting based on an optimum spacing between coded units entering the AAL2 multiplexer and the setting based on a worst case spacing of coded units.

5.4.2 Combining Coded Units into an AAL5 Frame

Example A single G.726 16-Kbps coder produces 10 bytes over 5 ms. An additional 11 bytes of overheads are added at the ATM interface. The coded voice signal requires one ATM cell for the transmission of each AAL5 frame. The cell is filled to 21 bytes—that is, 27 bytes are wasted. The total bandwidth required is 84.8 Kbps for a channel that originates at a bandwidth of 64 Kbps.

If the delay incurred in waiting for two coded units is tolerable, these will also fit into one ATM cell. After overheads have been added, the cell is filled to 31 bytes. The cell fill delay increases to 10 ms, but the average bandwidth requirement per channel reduces to 42.4 Kbps.

If three coded units are collected, the delay increases to 15 ms, but the average bandwidth per channel reduces to 28.27 Kbps and the cell is filled to 41 bytes. However, when four coded units are collected, these will no longer fit into one ATM cell, and the average bandwidth requirement per channel increases to 42.4 Kbps. As shown in Figure 5-2, the trend of the average bandwidth

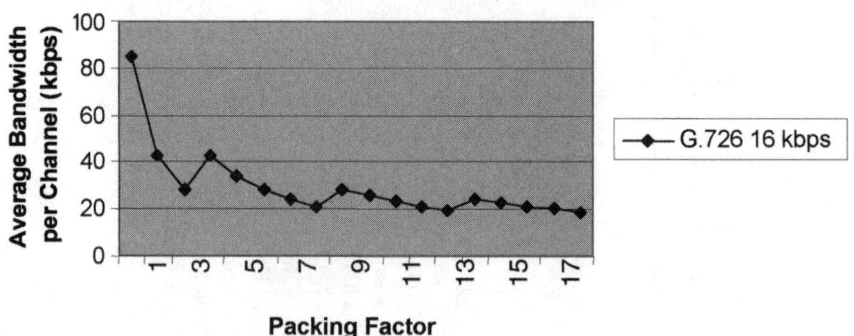

Figure 5-2 Increasing the packing factor of G.726 16 Kbps in an AAL5 frame

requirements is downward when the packing factor is increased, but the decrease is not even due to the granularity of the cell fill process—that is, a multiple of G.726 16-Kbps coder units does not fit exactly into an integer number of ATM cells.

Many coders offer a packing factor beyond four coding units. For example, G.726 16 Kbps offers a packing factor of up to 18. Each coder unit incurs a delay of 5 ms. Therefore, the result of choosing a packing factor of 18 implies a packing delay of 90 ms. As the packing factor is increased, the bandwidth gains are nonexistent or relatively small. For example, the bandwidth required for a packing factor of 8 is 21.2 Kbps per channel, and this is exactly the same as the bandwidth required for a packing factor of 16. Increasing the packing factor to 18 results in a small bandwidth reduction of 2.36 to 18.84 Kbps and hardly seems worth the extra wait of 10 ms in the cell fill process.

Table 5-4 shows the bandwidth requirements for a given number of channels and the packing factor settings for 10 coding methods using AAL5. It includes the FRF.11, AAL5, and ATM cell overhead.[4]

TABLE 5-4 Bandwidth Requirements for AAL5

Coder Type	Output Rate of Coder (bps)	Coder Unit (ms)	Packing Factor Range	Packing Factor Selected	Number of ATM Cells Required Incl. Overheads	Bandwidth Required Incl. Overheads (Kbps)
Clear Channel 64 Kbps	64,000	5	1 to 4	2	2	84.80
G.711 μ-law 64 Kbps	64,000	5	1 to 4	4	4	84.80
G.711 A-law 64 Kbps	64,000	5	1 to 4	3	3	84.80
G.723.1 5.3 Kbps	5,300	30	1 to 6	6	3	7.07
G.723.1 6.3 Kbps	6,300	30	1 to 7	1	1	14.13
G.726 16 Kbps	16,000	5	1 to 18	4	2	42.40
G.726 24 Kbps	24,000	5	1 to 15	1	1	84.80
G.726 32 Kbps	32,000	5	1 to 9	3	2	56.53
G.726 48 Kbps	40,000	5	1 to 7	7	4	48.46
G.729A 8 Kbps	8,000	10	1 to 13	10	3	12.72

[4] Formulas for the derivation of the tables in this book are given in Appendix H.

5.5 Comparison of AAL1, AAL2, and AAL5 Bandwidth

Figure 5-3 illustrates the bandwidth requirements for Clear Channel or G.711 [1] (A law or μ-law), where the PCM coded voice is not compressed, and for G.726 [3], where the voice is compressed to 16 Kbps using AAL5 and a packing factor of 1 to 4. The effect of using AAL2 is shown for comparison.

In all cases, use of AAL1 results in an average bandwidth requirement of 72.8 Kbps for each 64-Kbps channel received from the voice source. Table 5-5 compares the average bandwidth requirements of Clear Channel and 8 different methods of coding voice using AAL1, AAL2, and AAL5. The packing factor is varied between 1 and 4 for AAL5. In AAL2, it is assumed that the timer setting permits the insertion of sufficient coded units into the AAL frame to enable the ATM cell to be filled completely.

Figure 5-4 illustrates the comparison of the average bandwidth requirements per channel. A corresponding diagram of the delay incurred is shown in Figure 5-8.

5.6 Comparison of Delay in AAL1, AAL2, and AAL5

ITU-T Recommendation G.114 [22] discusses objectives for the limits on maximum one-way delay. This includes all processing, and propagation and queuing delay. Ideally, the total end-to-end delay between two interconnected telephones should be less than

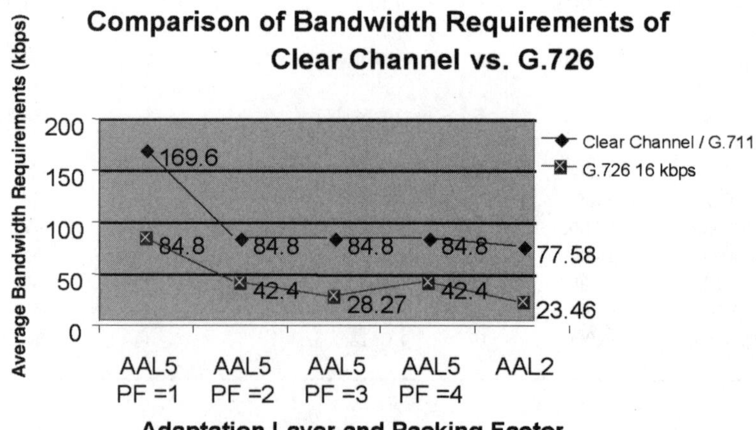

Figure 5-3 Comparison of average bandwidth per channel on compressed and uncompressed voice using Clear Channel/G.711 and G.726

TABLE 5-5 Average Bandwidth Requirements per Channel for Clear Channel and Eight Voice Coders[5]

Comparison of Average Bandwidth Requirements (Kbps) of Different Methods of Coding Voice						
Coder Type	AAL5 PF=1	AAL5 PF=2	AAL5 PF=3	AAL5 PF=4	AAL2	AAL1[a]
Clear Channel/ G.711	169.6	84.8	84.8	84.8	77.58	72.8
G.723.1 5.3 Kbps	14.13	14.13	9.42	7.07	6.88	72.8
G.723.1 6.3 Kbps	14.13	14.13	9.42	10.6	8.01	72.8
G.726 16 Kbps	84.8	42.4	28.27	42.4	23.46	72.8
G.726 24 Kbps	84.8	42.4	56.53	42.4	32.48	72.8
G.726 32 Kbps	84.8	84.8	56.53	42.4	41.5	72.8
G.726 40 Kbps	84.8	84.8	56.53	63.6	50.52	72.8
G.729A 8 Kbps	42.4	21.2	14.13	21.2	11.73	72.8

[a] All AAL1 results are based on an E1 basic consisting of 31 channels with no CAS.

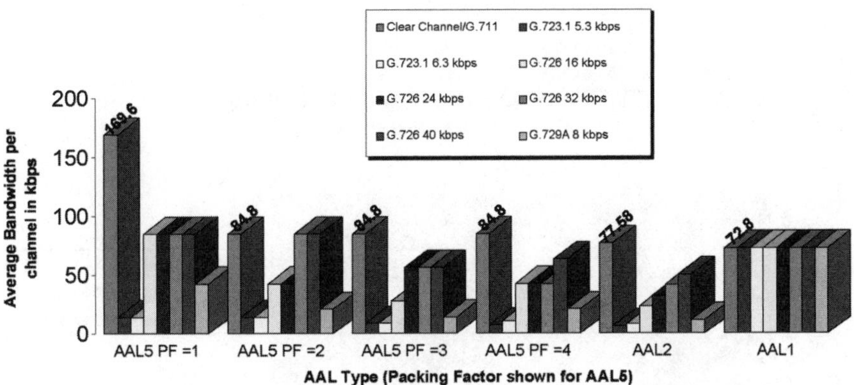

Figure 5-4 Comparison of AAL1/AAL2 and AAL5 bandwidth requirements

[5] Clear Channel and G.711 (μ-law or A-law) have the same bandwidth requirements and therefore are combined in a single row of results. Some coders offer packing factors higher than H.

150 ms, and echo cancellation[6] should be used for end-to-end delay that is greater than 16 ms. It is generally important to reduce delay wherever possible, including the edge of the network. Thus, any trade-offs between delay and bandwidth efficiency have to be looked at.

In Structured Circuit Emulation using AAL1, the delay incurred is dependent on the number of channels that are sampled and the fill level of the ATM cell. Figure 5-5 illustrates the average cell fill delay incurred when cells are filled to 47, 23, and 16 bytes, and the number of contributing voice channels is varied from 1 to 31.

As shown in Figure 5-5, the maximum delay incurred at full cell fill by a single voice channel is 6.375 ms. The minimum delay of circa 0.005 ms occurs when the maximum number of channels (31) are contributing to the filling of one AAL1 frame and/or the cell fill rate is extremely low, for example, 1 or 2 bytes.

Figure 5-6 compares the delay incurred for 30 voice sources using a Clear Channel/G.711 coder with different adaptation layers: AAL5 with a packing factor of 1 to 4 is compared with use of AAL2 (with the optimum timer setting) and AAL1 (using full cell fill).

Each channel using AAL1 is sampled a byte at a time, whereas the minimum sample size in AAL2 and AAL5 is one coder unit.

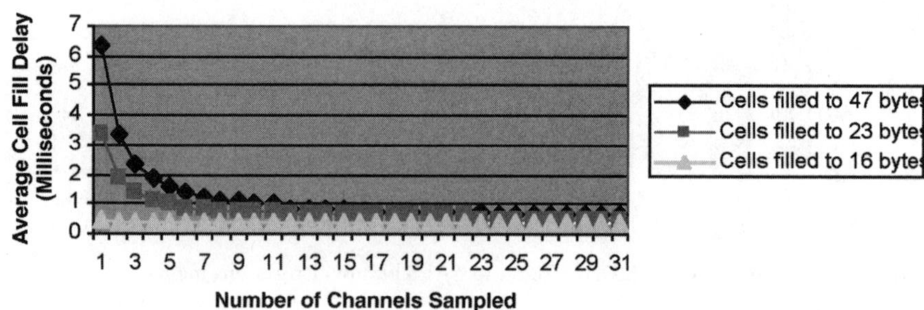

Figure 5-5 Average cell fill delay in AAL1 for full cell fill

[6] ITU-T Recommendations such as G.131, G.114, G.165, and G.168 contain information on the subject of echo cancellation. This book does not address the methods used to determine the need for, or the procedures for, applying echo cancellation.

Therefore, the minimum cell fill delay incurred by both AAL2 and AAL5 for one channel is the duration of one coder unit, as described in Table 5-2.

When several channels are contributing to the interface, the cell fill delay will depend on the number of channels. As channels are added, the delay decreases proportionately as the cell fill process is shared by the contributing channels. Figure 5-7 illustrates results for a G.726 16-Kbps coder, showing how the cell fill delay can reduce when AAL2 is used as additional channels are added.[7]

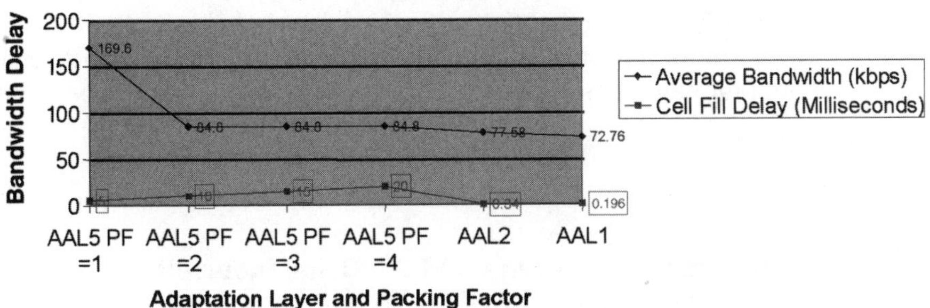

Figure 5-6 Comparison of average bandwidth per channel and cell fill delay for AAL1, AAL2, and AAL5

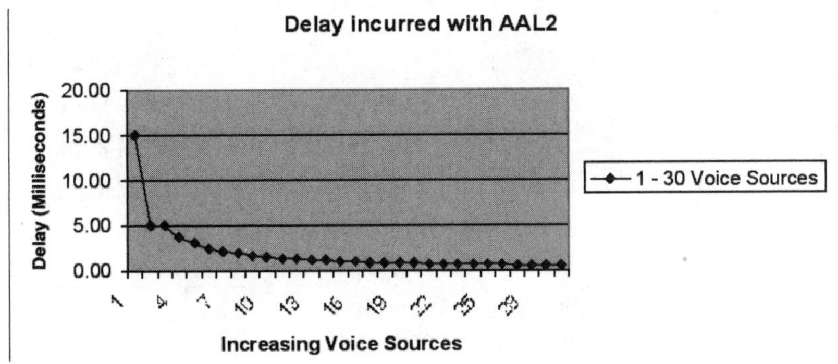

Figure 5-7 AAL2 cell fill delay incurred by 1 to 30 G.726 16-Kbps coded voice sources

---

[7] The results are for a 45-byte payload—that is, an AAL2 frame forms the payload of a single cell.

Initially, for one channel, each 5-ms coder unit is only 13 bytes long, and four coder units are needed for full cell fill. As channels are added, multiple coder units are ready more frequently and the cell fill delay reduces proportionately.

In the special case where there is only a single channel, the delay is simply a multiple of the duration of a coder unit in milliseconds multiplied by the number of coder units needed to fill the AAL2 frame. Both of these cases can be overridden by reducing the timer setting so that a frame/cell is sent as soon as a sample is placed in it, although the frame/cell may only be partially filled. In AAL5 where each channel has its own AAL interface, the selected cell fill configuration is the only factor that can reduce the cell fill delay—this would be a minimum of one coder unit in duration.

Figure 5-8 and Table 5-6 show the average cell fill delay (ms) incurred by different coders when there are 30 channels. Results are shown for a packing factor of 1 to 4 for AAL5. AAL2 results are shown for comparison, and it is assumed that the timer does not

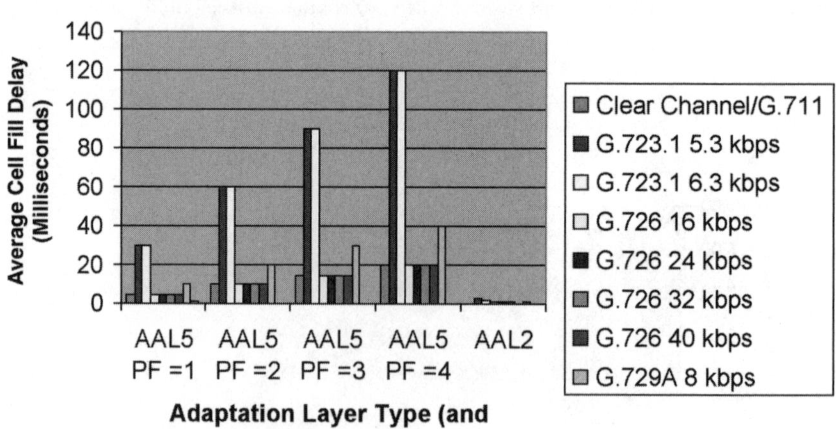

Figure 5-8 Comparison of average cell fill delay for 30 voice channels using different adaptation layers and packing factors

TABLE 5-6 Comparison of AAL1, AAL2, and AAL5 Average Cell Fill Delay Based on 30 Voice Channels

	Comparison of Cell Fill Delay (ms) for Different Adaptation Layers					
Coder Type	AAL5 PF =1	AAL5 PF =2	AAL5 PF =3	AAL5 PF =4	AAL2	AAL1
Clear Channel/G.711	5	10	15	20	0.34	0.196
G.723.1 5.3 Kbps	30	60	90	120	3	NA
G.723.1 6.3 Kbps	30	60	90	120	2	NA
G.726 16 Kbps	5	10	15	20	0.67	NA
G.726 24 Kbps	5	10	15	20	0.5	NA
G.726 32 Kbps	5	10	15	20	0.5	NA
G.726 40 Kbps	5	10	15	20	0.34	NA
G.729A 8 Kbps	10	20	30	40	1.34	NA

prevent full cell fill. AAL1 cell fill delay is always constant at 0.196 ms.

5.7 Diverse Bandwidth Requirements of a Single Voice Source

As illustrated in Figure 5-9, regardless of the type of adaptation layer framing being used, a single voice source may intermittently present several very different bandwidth requirements:

- A channel transporting uncompressed, non-silence-suppressed voice requires a Clear Channel at 84.8 Kbps (AAL5) at a constant

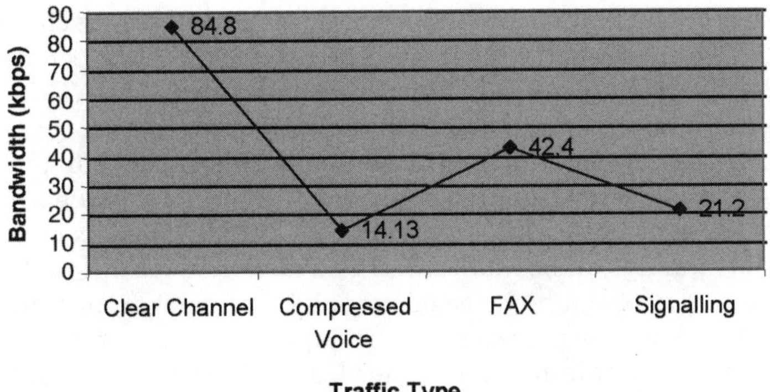

Figure 5-9 Intermittent bandwidth requirements of a single voice source

rate. Occasionally, a channel may need to switch from a compression coder to this mode, for example, when DTMF is being used.

- The same channel transmitting voice using a G729A compression coder (8 Kbps) with a packing factor of 6 will require 14.13 Kbps (encapsulated in AAL5, without silence suppression) at a variable rate. This low bit rate may result in unsatisfactory fax or modem performance.
- When fax or modem traffic is detected, a higher-rate bypass coder can be applied for transporting the modulated signal. For example, suitable bypass coders are G.711 μ-law or A-law, or G.726 32 Kbps. A channel using fax with a bypass coder G.726A (32 Kbps) and a packing factor of 6 (for AAL5) will require 42.4 Kbps at a constant rate.
- A channel using CAS will typically require 21.2 Kbps (AAL5) at a constant rate whether or not voice applications are being transmitted.

Additionally, the four traffic types do not behave in a similar way—that is, the arrival and duration of the traffic types vary. The voice is considered to be variable in arrival and duration. The signaling is constant in both. The fax is variable in arrival and constant in duration.

5.8 Silence Detection and Suppression

Additional bandwidth savings can be achieved in speech channels by not transmitting cells over the ATM network during periods of silence. Silence detection can identify silence within a conversation by monitoring the power level and time-varying characteristics of the signal in the ingress information path. Upon detection of silence, the transmission of speech data into the ATM network is suppressed for the duration of the silence.

Silence suppression/removal and compression coding are usually combined within the same software process. A silence indicator is sent to the other side to warn that silence has occurred so that timing can be maintained and so that a synthesized comfort noise signal can be inserted to replace the background noise that would have been present in the suppressed packets. In this way, packets containing silence need not be transmitted across the network and further bandwidth gain can be achieved.

5.9 Benefits of Silence Suppression

Silence suppression can normally be enabled/disabled for each individual voice channel. In order to understand the gains that can be made through not sending silence packets, it is important to understand generally how often silence occurs in a conversation. This was investigated by Brady [31] who showed that the average talkspurt lasts about 1.34 sec and the average silence spurt lasts about 1.67 sec, which means that approximately 55 percent of a conversation consists of silence. Therefore, on average, we could benefit from that scale of reduction in the bandwidth required by a call and the reduction applies in both directions.

However, in order to allow for a worst-case scenario where a number of sources are talking simultaneously, the scale of the reduction in the provisioned bandwidth allowed for due to silence suppression must be calculated carefully. The benefits of silence suppression include the following:

- Bandwidth reserved for a connection may be used temporarily by other connections during periods of silence. This technique is referred to as *Dynamic Bandwidth Allocation* (DBA). In particular, *Unspecified Bit Rate* (UBR) and *Available Bit Rate* (ABR) services may benefit from bandwidth made available due to DBA.
- Statistical multiplexing gains may be achieved across a group of channels, and hence across the connections carrying those channels when silence suppression is enabled on multiple channels within that group.

5.10 Silence Detection Overheads

The operation of silence suppression is normally independent for both directions of a channel. Usually, one party is silent while the other is talking, which causes data to be suppressed in one direction only. Silence suppression in both directions occurs during periods where both parties are silent. (Silence suppression software normally turns itself off when a fax or modem is detected.) To mask audible effects of silence suppression, the far side generates comfort noise in the egress direction during suppressed periods.

The silence detection process incurs some operational overheads. At the end of each talkspurt, when the silence is first detected, a short *Silence Indicator Packet* (SID) is added to the talkspurt data, which will slightly increase the length of the talkspurt. Some variance exists between different coders in the number of SIDs that are sent and this also depends on the level of background noise.

Additionally, bandwidth calculations also allow for a heartbeat packet of one cell, which is sent every 5 sec. This is a keepalive mechanism to detect if the ATM link is still alive. This cell would be sent during both talkspurts and silence.

Conclusions

- Commonly used compression coders vary in the rate of compression offered and can reduce a 64-Kbps voice channel to rates that vary down to a minimum of circa 5.3 Kbps.
- Using AAL1, which is a Constant Bit Rate (CBR) service, bandwidth requirements are not further reduced by compression or statistical multiplexing gains and are always increased by ATM framing overheads.
- When AAL2 and AAL5 are used, framing overheads, cell headers, and unused space in the ATM cells caused by partial filling will reduce the bandwidth gain offered by a compression coder. Nevertheless, the compression process is still very worthwhile.
- In the selection of a packing factor in AAL5 or the setting of a timer in AAL2, try to find a balance between reducing bandwidth requirements and keeping the cell fill delay times to a reasonable amount. You should determine if the delay incurred in waiting for additional coder units is tolerable; in order to do this, it is necessary to look at the budget for end-to-end delay across the whole connection.
- When AAL2 is applied, coder units from any channel may contribute to the same AAL interface. The delay in filling a cell is greatly reduced when several channels contribute. For example, AAL2 offers only slightly more delay at cell fill than AAL, when 30 channels are contributing coder units on a round-robin basis.
- AAL2 also offers the greatest bandwidth savings for all compressed voice, but the combining of voice samples from different channels onto the same adaptation layer interface requires that all voice connections from these sources are going to the same port at the exit point from the ATM network unless AAL2 subcell switching is offered, which is not often the case, at the time of this writing. Subcell switching is the ability to terminate an

AAL2 ATM Virtual Connection (VC), extract the AAL2 packets received on that VC, and forward those packets onto other ATM VCs when necessary, translating the packet Channel Identifiers (CIDs) to different values in the process. Without this ability, all of the channels that leave together from one AAL2 connection in a single VC must also arrive together in the same VC at the destination AAL2 linecard—that is, it is not possible for channels that leave together in the same VC to terminate at different AAL2 linecards.

- AAL5 offers the flexibility of a separate ATM VC for each channel where connections are not necessarily going to the same destination.
- Also, in some cases where there is a higher rate of compression, there may be very little difference between the bandwidth requirements using AAL5 and the bandwidth requirements with AAL2—for example, G.726 32-Kbps AAL5 (PF = 4) requires 42.4 Kbps and AAL2 requires 41.50 Kbps. However, the delay will be much higher with the AAL5 implementation.
- When interworking with AAL5 elsewhere in the network is desirable, and/or interworking with frame relay is a requirement, it may be possible to get close to the bandwidth savings offered by AAL2 by increasing the packing factor and using echo cancellation to reduce the effects of the increased delay.
- The application of silence suppression in AAL2 and AAL5 may further increase the bandwidth savings and reduce the difference between the bandwidth rates from AAL2 and AAL5, particularly when a higher rate of compression is used, for example, G.729A, which reduces voice from 64 to 8 Kbps.
- Subject to tolerable delay, it is desirable to strive for the lowest bandwidth requirement, which may be achieved when both silence suppression/removal and compression are applied to the voice channel.
- It is then difficult (if not impossible) to form traffic descriptors, which apply to the long-term steady state of the voice channel(s), when the intermittent uses of the channel (that is, voice, fax, DTMF, and signaling) are so variable in their bandwidth intensity.
- Therefore, a problem is identified as being the means of estimating the collective bandwidth requirements for a number of

channels at the point at which they are transmitting silence-suppressed and/or compressed voice, but in the presence of other voice-like services that may be using the channels intermittently.

- When different traffic types use the same channel, the traffic descriptors would have to be set to meet the highest bandwidth requirement (that is, the worst case), which would not necessarily reflect the parameters applicable for the average use of the channel during the majority of the time—for example, when the channel is being used for silence-suppressed and/or compressed voice.
- The use of the same channel for different traffic types complicates the possibility of using Call Admission Control (CAC), which is based on the long-term steady state of the channels as a means to setting the effective bandwidth and achieving statistical gain due to compression and silence suppression.
- Overbooking the port may be considered to be the only way of achieving statistical gain due to silence suppression and/or compression. There are two provisos in taking this approach: (1) the extent to which overbooking can be applied should be such that the performance parameters of the class of service continue to be guaranteed, and (2) the recommended bandwidth to be provisioned using this approach should be checked against the bandwidth that would have been provisioned if the CAC function was applied solely to the channels carrying voice.

Chapter

6

Quality of Service in an ATM Network

This chapter addresses the two main issues in preserving the *Quality of Service* (QoS) requirements of silence-suppressed and/or compressed voice channels.

- The first question concerns the selection of the maximum number of voice sources to allow in the system when a reduction is made in the provisioned bandwidth without degrading the QoS of the existing connections. The target *Cell Loss Ratio* (CLR) of the selected QoS class will reflect the probability that the egress buffer of the ingress line card could be exceeded, as illustrated in Figure 6-1.
- The second question concerns the maximum queuing time in a shared buffer for a predetermined number of sources and grade of service. It is assumed that the buffer is sized according to the service class of the traffic. The target *Cell Delay Variation* (CDV) will be affected by the size of the buffer and the rate at which the buffer queue is served.

6.1 Traffic Management in an ATM Network

An ATM network must be able to carry many types of traffic, including voice, each of which may have different requirements. This necessitates sophisticated traffic-management capabilities that must be able to control network congestion and make efficient use of network resources. The traffic-management functions

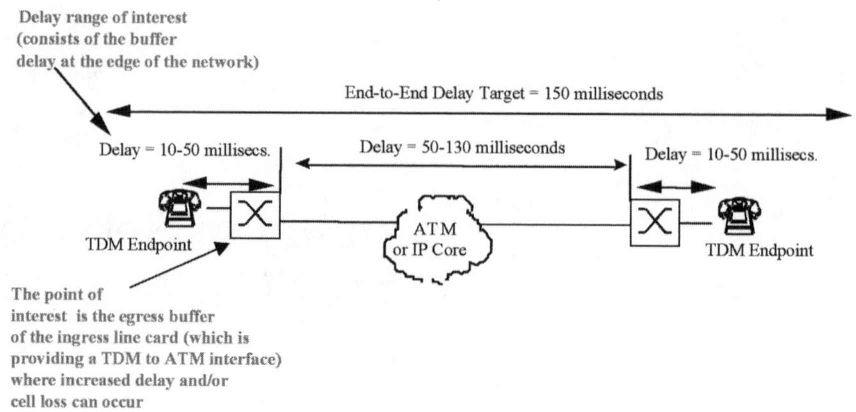

Figure 6-1 Potential delay and cell loss in the buffer at the point of entry to the network

must also be *ATM Adaptation Layer* (AAL) independent and must be able to support multiple QoS classes. Each QoS class creates a specific combination of bounds on performance values.

6.2 Service Categories and QoS Classes

Service categories[1] relate traffic parameters and QoS requirements, as shown in Figure 6-2. In general, routing, *Connection Admission Control* (CAC), and resource allocation functions are structured differently for the various service categories.

TM 4.0 [58] (the traffic-management specification of the ATM Forum's Technical Committee) specifies the following service categories:

- *Constant Bit Rate* (CBR)
- *Real-time Variable Bit Rate* (Rt-VBR)
- *Non-real-time Variable Bit Rate* (Nrt-VBR)
- *Available Bit Rate* (ABR)
- *Unspecified Bit Rate* (UBR)

TM 4.1 [59] adds a *Guaranteed Frame Rate* (GFR) service category, and a separate addendum [60] adds a UBR + *Minimum Cell Rate* (MCR) service category. GFR is intended for the transport of

[1]All service categories apply to both *Virtual Channel Connections* (VCCs) and *Virtual Path Connections* (VPCs).

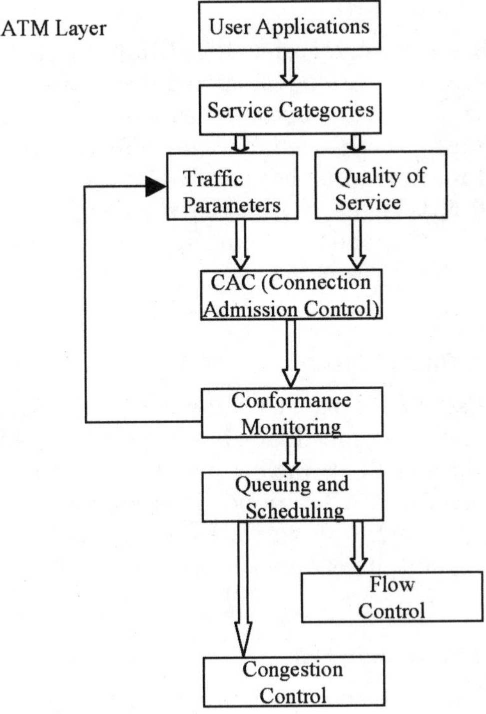

Figure 6-2 Elements of traffic management

data services such as the interconnection of IP routers over ATM. MCR is a parameter for a traffic descriptor, which describes the rate at which a source is always allowed to send.

6.2.1 Constant Bit Rate (CBR)

The CBR service category is intended for nonburst real-time applications and is typically used to support circuit emulation or private leased-line services.

CBR connections have the following:

- A fixed timing relationship between samples, meaning that there is a limit on the amount of CDV that can be experienced by a connection.
- A limit on the end-to-end delay; the *Cell Transfer Delay* (CTD) must be specified.
- A guaranteed low cell loss; the CLR must be specified.

6.2.2 Variable Bit Rate (VBR)

The VBR service categories, Rt-VBR and Nrt-VBR, are intended for bursty applications that require buffering. For example, Rt-VBR is typically used to support compressed voice, and Nrt-VBR is typically used for *Frame Relay* (FR) interworking.

The CLR target must be specified for connections in both the Rt-VBR and Nrt-VBR service categories. Only the Rt-VBR service category requires the additional specification of the CDV and CTD targets for the connection.

6.2.3 Available Bit Rate (ABR)

The ABR service category is intended for bursty applications utilizing some form of feedback-based rate control. ABR is typically used for data or file transfers over bandwidth unused by CBR or VBR connections. ABR connections do not require that CDV or CTD be specified. A quantitative value of CLR may be required (depending on the network).

6.2.4 Unspecified Bit Rate (UBR)

The UBR service category is intended for applications that have no CDV, CTD, or CLR requirements. UBR is typically used for e-mail or bulk transfers.

The TM 4.0 specifications [58] refer to UBR as *best effort*, and indicate that the *Peak Cell Rate* (PCR) and *Cell Delay Variation Tolerance* (CDVT) can be specified, but that these values cannot be subject to CAC and *Usage Parameter Control* (UPC) procedures.

6.3 ATM QoS Classes

Within a service category, there can be multiple QoS classes each with different traffic parameters. For example, a Rt-VBR service category could be divided into two QoS classes, Class I and Class II, where Class I offers a more stringent CDV traffic parameter than that required by Class II.

Provisional QoS class definitions and network performance objectives are defined in ITU-T Recommendation I.356 [64]. Four classes are defined: 1 (stringent), 2 (tolerant), 3 (bilevel), and U (no commitment to any parameters).

Each QoS class has a specific combination of bounds on the performance values. ITU-T Recommendation I.356 [64] includes guidance as to when each QoS class might be used, but it does not mandate the use of any particular QoS class in any particular context.

ITU-T Recommendation I.356 [64] differs from other ITU-T performance recommendations because the user has the option of requesting a different QoS for each new VPC and VCC; for some QoS classes and certain performance parameters, the ITU-T will not recommend any minimum level of quality.

6.4 ATM QoS Class Parameters

ATM layer QoS is measured by a set of parameters, which quantify the end-to-end network performance at the ATM layer. In ITU-T Recommendation I.356 [64], these parameters are referred to as *network performance parameters*.

Each QoS service class within a service category is usually associated with measurable network performance parameters. ATM Forum Traffic Management TM 4.0 [58] specifies that the QoS values can be specified individually, instead of being implicitly specified via QoS classes.

The following parameters are defined in ITU-T Recommendation I.356 [64]:

- *Cell Delay Variation* (CDV) (peak to peak)
- *Cell Transfer Delay* (CTD) (maximum)
- *Cell Loss Ratio* (CLR)
- *Cell Error Ratio* (CER)
- *Severely Errored Cell Block Ratio* (SECBR)
- *Cell Misinsertion Rate* (CMR)

A QoS class could contain, for example, the following performance parameters: maximum/mean CTD for the aggregate flow, CDV for the aggregate flow, and CLR for the aggregate flow. CDV, CTD, and CLR are all negotiated parameters, meaning that the actual values are determined by the physical network configuration. CER, SECBR, and CMR are not negotiated.

6.4.1 Cell Loss Ratios (CLRs) Within a QoS Class

Many factors can lead to network congestion, including the excessive bursting of traffic, overbooking of traffic, rerouting of connections, improper configuration of the ingress policing, and so on. Congestion causes increased cell buffering, which in turn causes delays and may lead to cell loss.

The CLR for a connection is defined as follows as per the ATM Forum TM 4.0 specification [58]:

$$\text{CLR} = \frac{\text{Lost Cells}}{\text{Transmitted Cells}} \tag{6-1}$$

Example A specified CLR of 1.0e-09 is a long-term statistical measure, meaning that for a connection that lasts forever, approximately one cell will be lost in every 1,000,000,000 cells, or for a large number of connections from the same source, the average CLR will be 1 in 1,000,000,000. None of these two explanations can tell the user about the CLR that a specific connection will actually experience in its entire life. Because cell losses tend to occur in clusters, many busy hours will have no cell loss, while an occasional busy hour will experience a CLR much higher than 1 in 1.0e-09.

The probability that cell loss occurs during a given period is still a useful and relevant measure of QoS, and it is possible to define two CLR parameters: CLR_{0+1}, which applies to the aggregate cell stream, and CLR_0, which applies to the higher-priority stream.

6.4.2 Cell Transfer Delay (CTD) Within a QoS Class

It is expected that other performance parameters as well as the CLR will apply to the aggregate cell flow of the ATM connection. ATM CBR and Rt-VBR service classes require the estimates of the CLR and CTD to be stated.

CTD is defined as the elapsed time between a cell exit event at the source *User Network Interface* (UNI) and the corresponding cell entry event at the destination UNI for a particular connection.

Some components of the delay are constant (fixed), and some are variable. The components of the fixed delay include propagation delay through the physical media, delays induced by the transmission system, and fixed components of ATM node processing delay.

The total ATM node processing delay encountered by a cell is comprised of the total delays encountered at different points in the network, for example, during queuing, adaptation, and switching.

Estimates of CTD are provided for each class of traffic, such as high priority and low priority. CDV is introduced by buffering and cell scheduling.

6.4.3 Cell Delay Variation (CDV) Within a QoS Class

CDV is a measure of the jitter (positive and negative) in the cell interdeparture pattern of a given connection with respect to its interarrival pattern. This performance measure is illustrated in Figure 6-3.

Two methods are defined by the ATM Forum for the measurement of CDV and are summarized here:[2]

- **One-point CDV** This describes the variability in the pattern of cell arrival events observed at a single measurement point with reference to the negotiated *peak rate*, 1/T. The one-point CDV for cell k (y_k) at a measurement point is defined as the difference between the cell's reference arrival time (c_k) and actual arrival time (a_k) at the measurement point. Thus, $y_k = c_k - a_k$.
- **Two-point CDV** This describes the variability in the pattern of cell arrival events observed at the output of a connection portion (MP2) with reference to the pattern of the corresponding events observed at the input to the connection portion (MP1). The two-point CDV for cell k (v_k) between two measurement points (MP1

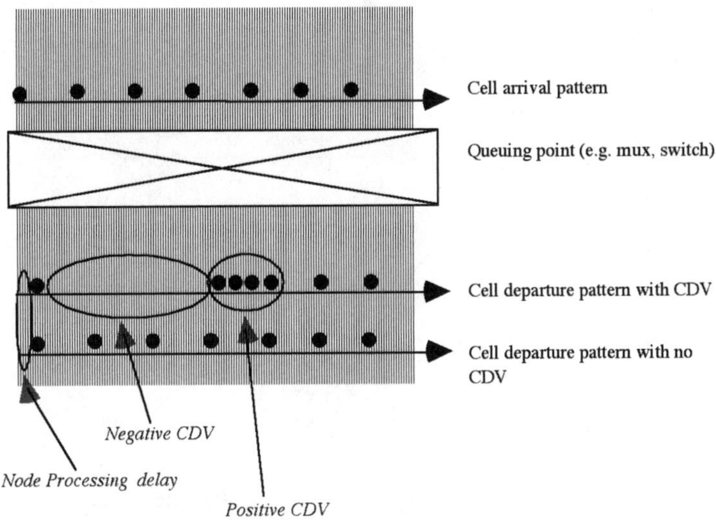

Figure 6-3 CDVT

[2] Refer to ATM Forum TM 4.0 [58] for complete details of the definitions.

and MP2) is the difference between the absolute CTD of cell k (x_k) between the two MPs and a defined reference CTD ($d_{1,2}$) between MP1 and MP2. Thus, $v_k = x_k - d_{1,2}$.

6.4.4 End-to-End CTD and CDV

Two end-to-end delay parameter objectives are negotiated: the maximum CTD and the peak-to-peak CDV. Figure 6-4 [58] illustrates how the probability density function of CTD is related to the parameters for peak-to-peak CDV and maximum CTD, which are specified for connections in the CBR and Rt-VBR service categories.

The probability density function of the CTD in CBR and Rt-VBR is right-tailed with a fixed minimum delay and an upper bound expressed as a quantile measure. The term *peak-to-peak* refers to the difference between the best and worst case of CTD, where the best case is equal to the fixed delay and the worst case is equal to a value likely to be exceeded with a probability no greater than (α).

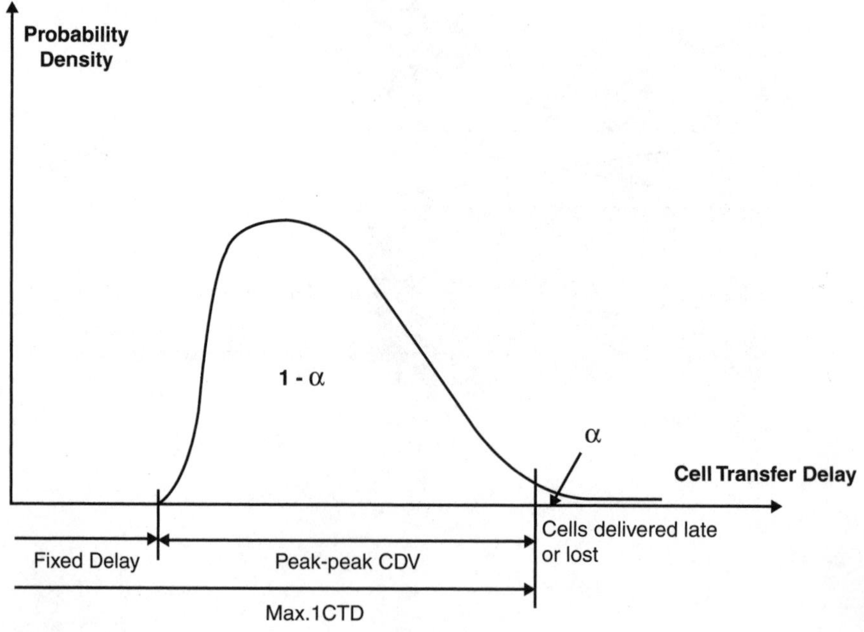

Figure 6-4 CTD probability density function [58]

6.5 Relationship Between Load, CLR, and CDV in a Single Shared Buffer

Consider the operation of one of the physical links that support a specific ATM connection [64]. All of the cells that are intended to pass through this physical link will be held in a buffer that absorbs momentary surpluses of cells until they are either transmitted over the link or until this buffer overflows with the resultant loss of some cells. The cells that are intended to pass through this physical link are provided by both the specific voice connection under consideration and other ATM connections that share this link, and all of these cells combine to establish the link's offered load, which may be characterized by a utilization factor $P_{offered}$.

Any cell arriving at this buffer experiences a random waiting time, *W*, before it reaches the link and is transmitted. Figure 6-5 illustrates this situation, together with some representative probability density functions for *W* [64].

With a sufficiently high value of offered load, characterized in Figure 6-5 by ρ_{HI}, the tail of the probability density function will place a significant amount of weight beyond the buffer capacity *B*, as measured in cell interval times.

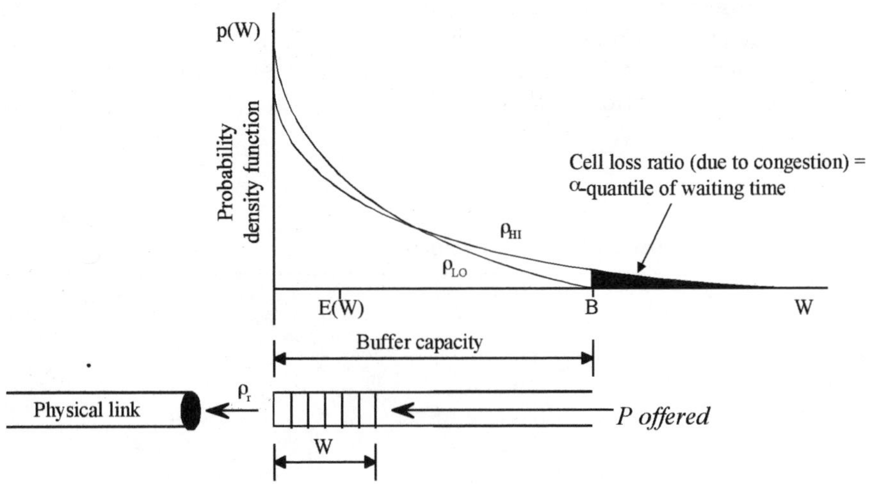

E(W) = Expectation of Waiting Time = $\int x f(x)\, dx$

Figure 6-5 Illustration of random waiting time [64]

The area under this curve can be interpreted as the CLR due to congestion. If the buffer were made larger, these cells would not overflow and the shaded area would then represent an upper quantile of cell waiting time. The maximum waiting time for a cell in this buffer occurs when the cell in question occupies the final available cell space. Thus, the maximum delay variation attributable to this buffer is controlled by the buffer size.

With a lower value of offered load, characterized in Figure 6-5 by ρ_{LO}, the tail of the probability density function will place less weight beyond B, thereby reducing the resulting value of CLR.

6.6 QoS Targets

Table 6-1 illustrates how a service category may be divided into several ATM QoS classes, each with varying QoS parameters. For example, the Rt-VBR service category could be divided into two classes, Class A and Class B, where a more stringent CDV target is applied to traffic in Class A. The Nrt-VBR service category is illustrated here as being divided into three classes: Class A, B, and C, which are differentiated by the CLR target. Additional examples to QoS parameters will refer to these sample classes as given in Table 6-1.

One (or more) of QoS parameter values may be offered on a per-connection basis, depending on the number of related performance objectives supported by the network. To support different performance values, the connection must either be rerouted or implementation-specific mechanisms within the network must be utilized.

TABLE 6-1 QoS Targets for Service Categories

Service Category	CLR	Target CDV (microsecs)
CONTROL	1.0e-10	N/A
CBR	1.0e-10	250
RT-VBR Class A	1.0e-09	250
RT-VBR Class B	1.0e-09	2500
NRT-VBR Class A	1.0e-07	N/A
NRT-VBR Class B	1.0e-06	N/A
NRT-VBR Class C	1.0e-05	N/A

6.7 Traffic-Descriptor Parameters

Traffic-descriptor parameters specify an inherent characteristic of a connection, which may consist of a single traffic source or several multiplexed traffic sources. Figure 6-6 illustrates the relationship between the traffic descriptors, which include the following:

- **Line rate** The capacity of the trunk carrying the connection. Several connections may be carried over the same trunk.
- ***Peak Cell Rate* (PCR)** Represents the peak emission rate of the sources(s). This is a cell rate, which the connection may exceed in the short term, but must conform to in the long term.
- ***Sustainable Cell Rate* (SCR)** Represents the theoretical long-term average rate of the connection.
- ***Maximum Burst Size* (MBS)** Represents the burstiness factor of a particular connection. MBS specifies the maximum number of cells that can be sent back to back at PCR, while still maintaining compliance with SCR.
- **Effective bandwidth** Represents the bandwidth allocated to the connection by the CAC algorithm within the ATM switch. This is further explained in Section 6.8.

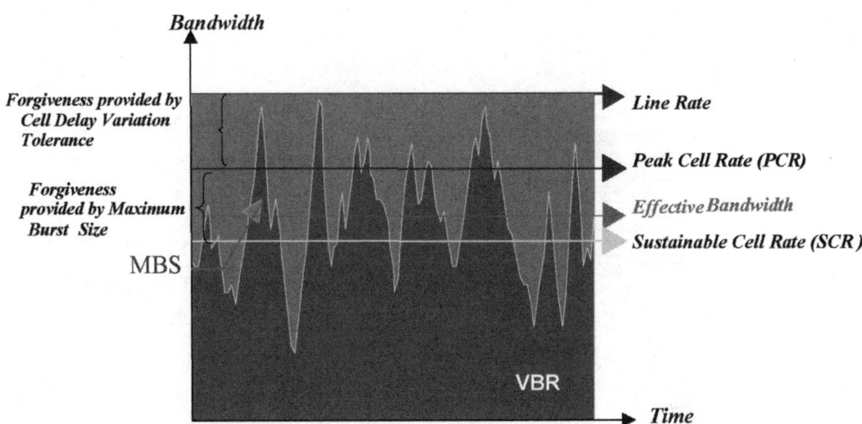

Figure 6-6 The relationship between traffic descriptors for a VBR connection

- ***Cell Delay Variation Tolerance*** **(CDVT)** Is used to configure the fill level of the buffer and is not taken into account by the CAC.

6.8 Call Admission Control (CAC) Algorithm

The CAC function allocates bandwidth to connections to provide the necessary CLR and CDV commitments for CBR, Rt-VBR, and Nrt-VBR service categories. For the UBR + MCR and ABR service categories, MCR is allocated by the CAC. As illustrated in Figure 6-7, CAC also determines whether a connection request is admitted or denied. The connection request defines the source traffic-descriptor parameters and the requested QoS class. The capacity checking function is performed when an attempt is made to add a connection, change the service category, change the traffic descriptor, or change the policing attributes for an existing connection.

6.8.1 Steady State Queue Behavior

When a queuing system is first put into operation, and for some time afterwards, the number in the queue and in service depends strongly upon the initial conditions (such as the number of cus-

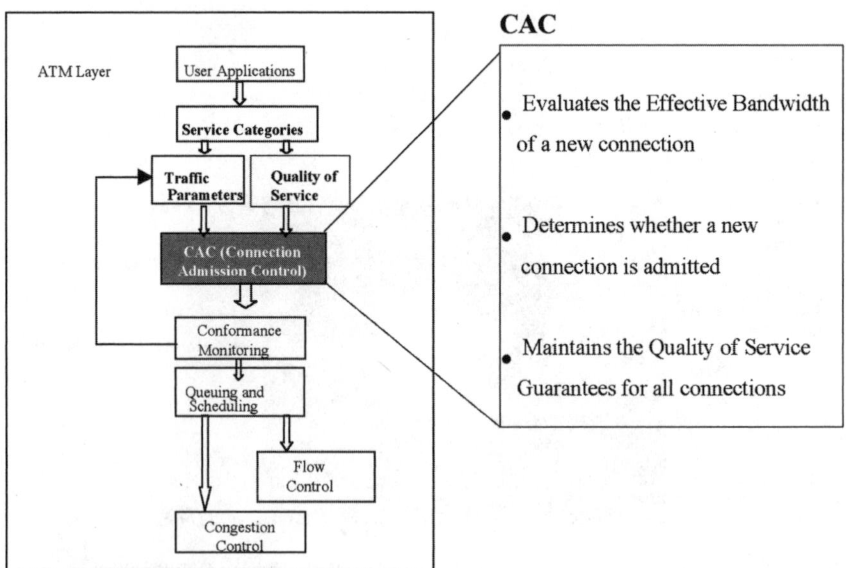

Figure 6-7 The role of CAC

tomers queued up waiting for the system to go into operation) and how long the system has been in operation (that is, the time parameter t).

The system is said to be in the *transient* state. However, after the system has been in operation for a long time, the influences of the initial conditions and of the time since startup have "damped out," and the number of customers in the system and in the queue becomes independent of time. The system is in the *steady* state.

Currently, proposed CAC algorithms or bandwidth allocation policies fall into two categories: one is the effective bandwidth approach and the other is based on real-time system evaluations. Both types of CAC algorithms are based on assumptions associated with a system's steady state. QoS metric specifications are also based on a system's steady state behavior.

6.8.2 Defining the Effective Bandwidth

The *effective bandwidth*, as shown in Figure 6-6, is a natural measure of a connection's bandwidth requirement relative to the QoS constraints, for example, delay and /or cell loss. The CAC works by virtually assigning each connection its effective bandwidth and rejecting a connection request when the remaining capacity is less than the connection's effective bandwidth.

The effective bandwidth calculation takes into account resources (bandwidth and buffers) and QoS parameters. In general, the same source with more stringent QoS requirements will demand a larger value of effective bandwidth from the network. The calculation of effective bandwidth requires PCR, SCR, and MBS inputs. It has been proven that the ratio of MBS to buffer size strongly influences the statistical gain when CAC is applied.

6.9 Overbooking

Application of the CAC may achieve some level of statistical gain. Additional statistical gain can be achieved by overbooking by taking into account the actual level of activity of the multiplexed connections. Overbooking enables network providers to achieve statistical multiplexing gain by having the sum of the allocated bandwidth of all virtual connections on a link exceed the physical bandwidth of the link.

Overbooking is based on the assumption that not all users will be using the network at the same time. However, where overbooking is configured on a congestion point for a specified service category, some of the service category performance objectives may not be guaranteed.

Conclusions

- All ATM service categories require adherence to a CLR target. The VBR service categories are of most interest here because these permit statistical gain in bandwidth when a number of connections are provisioned between two points. The CBR and Rt-VBR service categories require adherence to both CTD and CDV targets in addition to adherence to a CLR target.
- Both the CLR and CDV performance statistics are related to the size of the buffer.
- With respect to achieving the CLR target, there are two risks in reducing the provisioned bandwidth due to savings through silence suppression:
  - The risk of underestimating the mean length of a talkspurt that determines the total bandwidth requirements.
  - The risk of underestimating the number of sources that are likely to be in a talkspurt at the same time.
- The aim here is to provide a scaling factor for reducing the bandwidth according to a risk factor. In other words, the more aggressive the reduction made in bandwidth due to silence suppression, the higher the risk of cell loss. It should be possible to quantify that risk and show that the CLR of the selected QoS class can be maintained when overbooking is carefully scaled.
- The Call Admission Control (CAC) function within an ATM switch calculates the long-term effective bandwidth for one or several connections based on the QoS class and inputs of the traffic descriptors. PCR, SCR, and MBS must be input to the CAC when a VBR QoS class is selected.
- In the presence of silence suppression, the PCR traffic descriptor will be equivalent to a worst-case bandwidth requirement defined by the highest number of voice channels that are expected to be in a talkspurt concurrently. Some network managers may prefer to remain on the side of caution and size PCR according to the possibility that all channels will be in a talkspurt concurrently.
- In the presence of silence suppression, the SCR traffic descriptor will represent the average case defined by the average number of voice channels that are expected to be in a talkspurt concurrently.

- In the presence of silence suppression, the MBS traffic descriptor will represent the size of a burst of cells defined by the worst-case bandwidth requirement—that is, the highest number of voice channels that are expected to be in a talkspurt concurrently throughout the duration of the estimated talkspurt length.
- The sizing of PCR, SCR, and MBS will therefore depend on a number of factors:
  - The number of channels being combined together at the same interface.
  - The type of ATM Adaptation Layer (AAL) being used to encapsulate the coded units.
  - The type of compression coder being used to compress each voice channel.
  - The configured packing factor (in the case of AAL5) or the timer setting (in the case of AAL2).
  - The number of channels expected to be in a talkspurt concurrently.
  - The estimated length of an average talkspurt.
  - The estimated duration of the overlap of sources in a talkspurt concurrently.
- The aim, therefore, is to accurately characterize the traffic descriptors—PCR, SCR, and MBS—of silence-suppressed and compressed voice channels and to determine the extent to which overbooking can be applied.

Chapter

7

The Voice Model

This chapter seeks to define a representative model of voice traffic. In order to build an accurate voice model, it is necessary to select one or several appropriate levels at which to characterize the voice traffic. Cosmas et al. [26] recognized five resolutions in time for the characterization of ATM traffic:

- **Calendar level** Describing the daily, weekly, and seasonal traffic variations of a traffic source
- **Connection level** Describing the behavior of a traffic source on a virtual connection basis
- **Dialogue level** Describing the interaction between voice agents at both ends of the connection
- **Burst level** Describing the on/off characteristics of the cell generation process.
- **Cell level** Describing the behavior of cell generation at the lowest level

Burst level modeling reflects the time scale typical of an on/off source activity period [63]. Cell level modeling provides a corresponding bit rate. Hence, the last two of these five levels are selected as appropriate for this research, namely burst level and cell level. Their context is further described as follows:

- **At burst level (in talkspurts)** A bulk arrival of talkspurts at the buffer of the ingress linecard forms a burst. The size of a burst depends on three factors:
  - The estimated average length of a talkspurt
  - The frequency of the arrival of talkspurts from one or several voice sources sharing the connection
  - The duration for which talkspurts overlap when more than one voice source contributes to the same connection

  All of these factors determine the size of the burst in terms of talkspurts. The size of the burst is described by the *Maximum Burst Size* (MBS) parameter in *Call Admission Control* (CAC) procedures. Modeling at the burst level will establish an estimate for the MBS, which is normally described in units of cells. Translating a burst size into bandwidth, in order to obtain the maximum bit rate, will provide an estimate for the peak cell rate (PCR).

- **At cell level** The actual number of cells contained within such a burst of talkspurts depends on the following:
  - The selected adaptation layer type
  - The selected coder type

  Measurement at both the talkspurt level and cell level will be used to find the occupancy of the buffer and the delay incurred in waiting in the buffer.

7.1 Characterization of Voice

Figure 7-1 illustrates how a voice source alternates between active (ON) and inactive (OFF) states. During the ON state, a talkspurt has occurred; during the OFF state, a period of silence is taking place.

Brady's [31] work in the field of voice profiling is well respected and very applicable. The estimates of talkspurt and silence provided by Brady are used here as the basis of the calculations of talkspurt length.

The talkspurt and silence periods are exponentially distributed with mean $1/\beta$ and $1/\alpha$, respectively. The average talkspurt period is represented by $\beta = 1.34$ sec, and the average silence period is represented by $\alpha = 1.67$ sec [31]. The transition from talk to silence therefore occurs with the rate $1/\beta = 0.7463$ sec, and the transition from silence to talk occurs with the rate $1/\alpha = 0.5988$ sec.

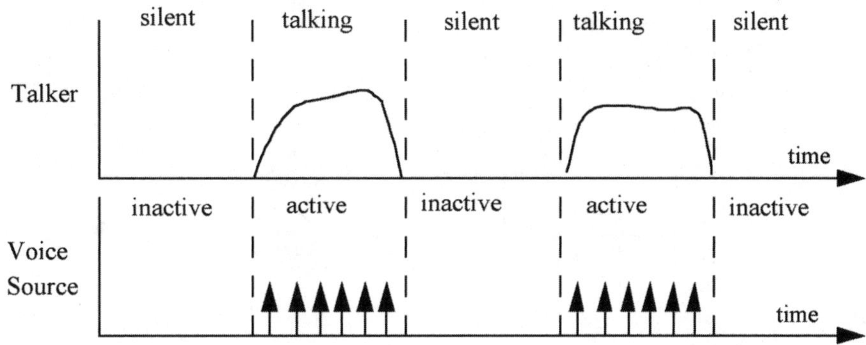

Figure 7-1 Voice behavior

7.2 Uniform Arrival and Service Model

Talkspurts arrive at a switch from N information sources, which independently and asynchronously alternate between the talkspurt and the silence state, as shown in Figure 7-2. Both the talkspurt and the silence periods are assumed to be exponentially distributed for each source. A common buffer queues the arriving cells, and a server of constant capacity empties the buffer at a uniform rate. The queuing model shown in Figure 7-3 represents a *Uniform Arrival and Service* (UAS) model.[1]

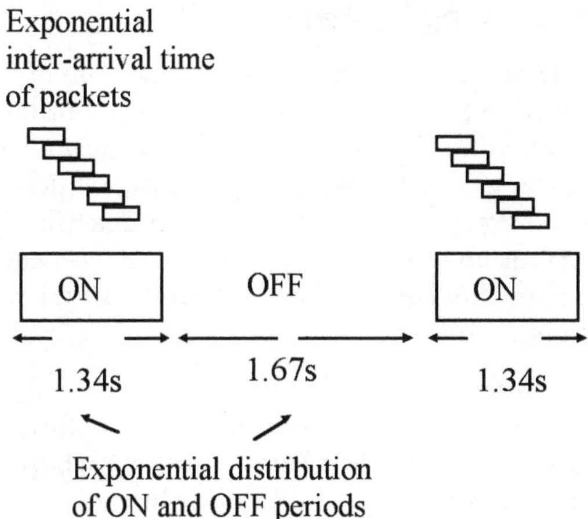

Figure 7-2 ON and OFF voice spurts

[1] See references Anick, Mitra, and Sondhi [24], and Daigle and Langford [27] for examples of the UAS model.

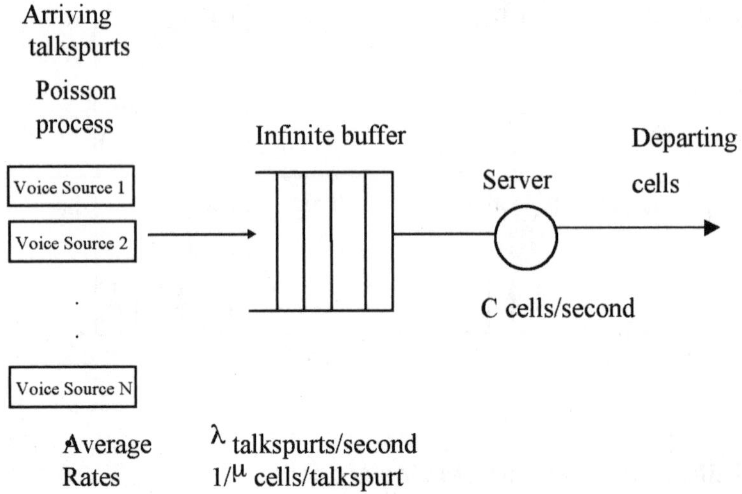

Figure 7-3 A UAS model for voice

Figure 7-3 illustrates how a buffer is drained at a rate of *C* cells per second. Shaping can be applied to determine the removal rate of cells from the buffer in order to enforce the traffic descriptors of a virtual connection. The effects of shaping are further described in Section 7.5.2.

7.3 Notation of Queuing Systems

The total time in a queuing system is the sum of the waiting time and the service time. In order to discuss queuing systems, a simple shorthand notation is used for describing queues. This involves the three-component description A/B/m, which denotes an m-server queuing system where A and B describe the interarrival time distribution and service distribution, respectively.

The convention for describing different arrival and servicing processes is as follows:

- **Memoryless (M)** This process appears where the process generating the occurrence of an event is fully independent of any other event. This is a Poisson process where the interarrival distribution is an exponential distribution.
- **General (G)** This process signifies any distribution.
- **Deterministic (D)** This process appears where the arrival time or service time is fixed.

For example, the M/M/1 queue describes a Markovian (memoryless) distribution with exponential interarrival times, exponential service times, and a single server.

7.4 Queue Emptying

Typically, a queue holds traffic from connections of similar *Quality of Service* (QoS) requirements, and an arbitration algorithm operates between queues. The modeling carried out in this research assumes that all sources contributing to the same connection use the same adaptation layer, belong to the same QoS class, and join the same queue at the egress buffer of the ingress linecard.

The maximum waiting time for a cell in the queue occurs when the cell in question occupies the final available cell space. Thus, the maximum delay variation attributable to this buffer is controlled by the buffer size and the rate at which the buffer is emptied.

7.5 Stochastic Modeling of a Single Queue in an ATM Switch

Several stochastic models have been proposed for the queue, and most of these models tend to be mathematically complex.

The queuing system with Poisson arrivals and deterministic service time is M/D/1. It is a simple special case of the M/G/1 and was studied by Erlang in 1909.

The M/D/1 queue is an appropriate model of an ATM multiplexer queue if we can assume that the cells arrive as a Poisson process. On the other hand, the M/D/1 model gives far too large tail probabilities for the queue length distribution in a heavily loaded system with periodic inputs [63].

The M/D/1 queue might reasonably describe a situation where the rate of the considered connection is small compared to the total input rate to the queue and the traffic feeding the queue is "smooth."

However, as shown by Sexton and Reid [49], a good understanding of buffering at the burst scale can be achieved by examining the restricted case where the full bandwidth is also allocated to a burst. This can be modeled by considering a simple M/M/1 queue where the arrival of bursts is modeled as a Poisson process, and the distribution of the length of the burst is modeled by the exponential distribution. The M/M/1 queue was also employed by Kelly et al. [46] to obtain straightforward analytical forms for system sizing. The probability of a number of bursts in the queue conforms to the standard state transition matrix for an M/M/1 queue.

The fill of the buffer depends on the number of bursts in the buffer and the size of those bursts. As it is known that the size of each burst is exponentially distributed, the distribution of n bursts will be the n-fold convolution of this exponential distribution.

The probability that a number of sources are simultaneously in a talkspurt leads to the sizing of the MBS and a corresponding buffer size—that is, for a given loss probability (related to the CLR target), the buffer size must be proportional to the mean burst size.

The probability that the buffer is filled to time t is the sum of the probability distribution across n talkspurts that make up a burst. The probability that the MBS will be exceeded is the same as the probability that a buffer of fixed size will overflow and cause loss.

Note that the M/D/1 system with constant service time has half the average waiting time of the M/M/1 system [47, page 191]. In fact, to achieve the same cell loss probability as the M/D/1 model of cell-scale buffering, the buffer must be increased by a factor of twice the mean burst size in cells [49].

7.5.1 Modeling at the Talkspurt Level

A burst consists of one or several overlapping talkspurts—that is, from one or several voice sources that arrive together. Each talkspurt may be further divided into several smaller groups of cells, with each small group comprising of an adaptation layer frame's worth of cells.

Example A G.726 coder at 32 Kbps provides a coder unit of 5 ms. Using AAL5 and a packing factor of 3 causes three coder units to be packed into one AAL5 frame. The three 5-ms samples of the compressed voice will be packed into the AAL5 frame, and the overheads incurred by the type of the adaptation layer selected will be added to the frame. In the AAL5 case investigated (using FRF.11 and FRF.8), such a frame would be transmitted using two ATM cells. If six samples are packed into one AAL5 frame (and the delay is tolerable), then three ATM cells will be required.

Assuming the packing factor of 3, as described in the first example, 180 cells will be needed to transmit the entire talkspurt, if the talkspurt length is estimated at an average 1.34 sec. The 180 cells will occur as 90 *minibursts* of 2 cells where each miniburst is an adaptation layer frame's worth of cells.

As shown in Figure 7-4, it is important to recognize that measurement of a bulk arrival of talkspurts is insufficient. In order to

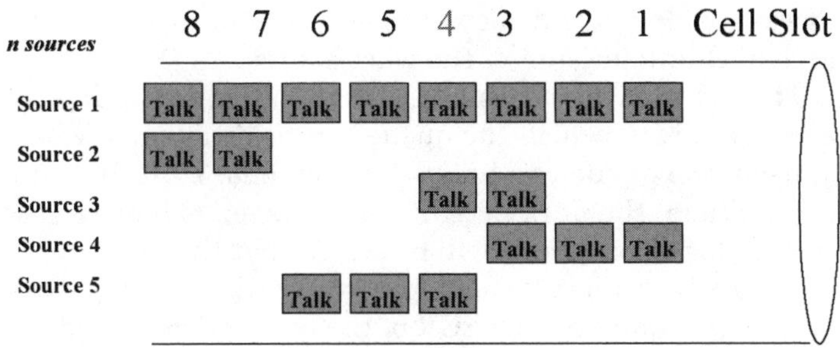

Figure 7-4 Bunching of cells in a talkspurt

understand the MBS, the overlap of talkspurts and the period of time for which this could occur must also be considered.

For example, in the simple example in Figure 7-4, 18 talkspurts are delivered over 8 talkspurt slots, resulting in an average cell rate of 2.25 talkspurts per talkspurt slot. In reality, the peak talkspurt rate is 3 talkspurts, occurring in slots 3 and 4.

The size of the talkspurts depends on the following:

- The estimated length of a talkspurt
- The adaptation layer type
- The coder type and duration of a coder unit, and hence, the number of cells carrying coded unit(s)

7.5.2 The Effects of Shaping

The objective is to discover the optimum setting for the traffic descriptors in the presence of compression and silence detection. Shaping will then be applied to ensure conformance to those traffic descriptors over time to ensure steady state behavior.

Shaping [64] is a process of equally spacing cells so that the intercell gap conforms to the traffic descriptors, that is, the desirable gap as determined by 1/PCR and 1/SCR.

A switch may apply shaping to the cells to ensure that they meet the requirements of the policing algorithm[2] (*Generic Cell Rate Algorithm* [GCRA]).[3] The effect of shaping is to incur a small

[2] Shaping may be applied whether policing is turned on or off.

[3] The GCRA is used to define conformance with respect to the traffic contract of the connection. For each cell arrival, the GCRA determines whether the cell conforms to the traffic contract [65, Annex A].

increased delay, and more importantly, to destroy some of the arrival characteristics of the voice source.

The delay incurred as a result of shaping will depend on the average rate at which the queue can be drained. If VBR shaping is used, the queue can be drained at some PCR for a number of cells. When the MBS credits have been exhausted, then the remainder of the queue will be serviced at the SCR.

Typically, connections should run without delay building up behind the shapers. The SCR of the connection should be at least large enough to keep up with the traffic generated by the average number of sources that are expected to be in a talkspurt at the same time.

7.6 Sizing the Buffer

The statistics of the voice sources and the traffic descriptors chosen will have a direct effect on the size of a burst of traffic that can be accommodated, assuming that shaping and/or policing are to be applied to enforce the traffic descriptors. A connection should be engineered to encounter a certain maximum delay build-up based on the maximum time it would take to drain out a bulge of traffic behind the shaper.

The aim is to discover the number of channels that can be carried by a link without exceeding the predicted MBS. The probability of cell loss caused by buffer overflow is expressed by the CLR target.

7.6.1 Meeting a Delay Target

If a queue is modeling a buffer and the buffer can hold a maximum of B cells to fit within a delay constraint, then the probability of cell loss can be found by determining the probabilities of all the fill states that are greater than B. This probability is the same as the summed probabilities of all the states of the MBS greater than 1.34 sec worth of cells.

A further aim can be expressed as, "Determine the traffic descriptors PCR, SCR, and MBS to be used to control the traffic shaping function, such that the probability that any cell passing through the traffic-shaping function incurs a delay greater than t is less than p."

Figure 7-5 illustrates how several voice channels accessing an ingress TDM-to-ATM linecard are multiplexed into a VC that is shaped and/or policed to ensure that the traffic descriptors are adhered to.

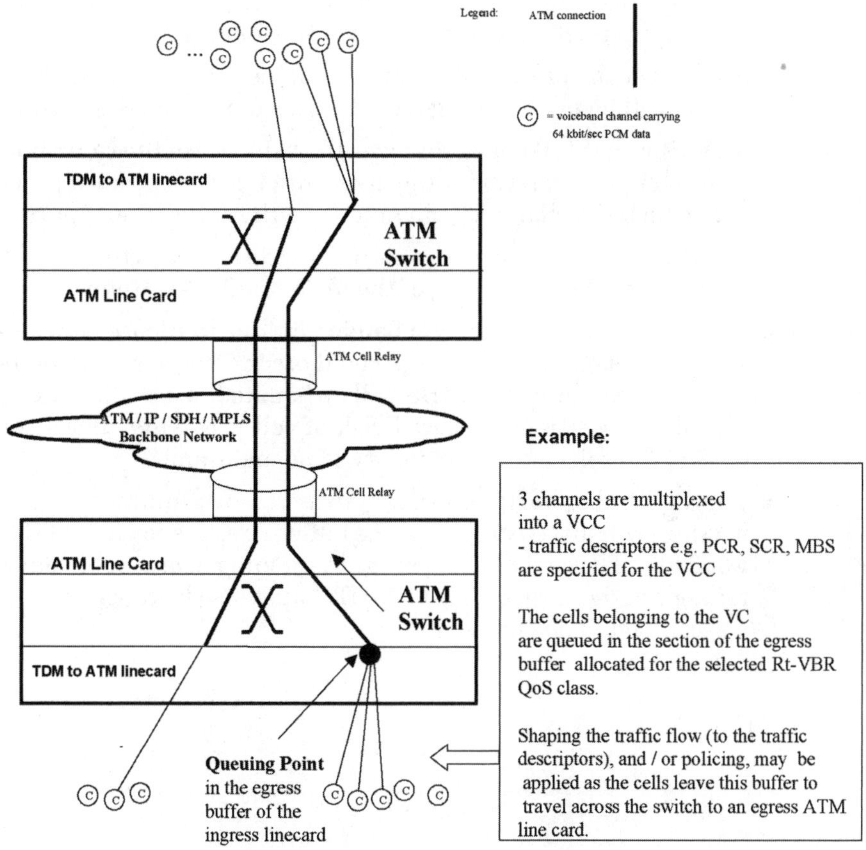

Figure 7-5 Voice traffic queuing at the ingress linecard

Conclusions

- Modeling will operate at two levels: the talkspurt and cell level.
- The potential for all contributing sources to be concurrently in a talkspurt will provide a measure of the Maximum Burst Size (MBS) parameter. The burst will be sized at a number of cells, which will depend on the number of contributing sources aggregated and the selected adaptation layer type and packing factor (AAL5) or timer setting (AAL2).

- In order to establish an appropriate setting for the MBS traffic descriptor, it will be necessary to do the following:
  - Discover the probability that the number of aggregated channels will be in an average-length talkspurt at the same time.
  - During that talkspurt, discover the behavior of the contributing channels in terms of the distribution of the expected times at which the channels join and leave the state of talkspurt.
- The degree to which concurrent talkspurts are expected to overlap will provide a measure of the MBS traffic descriptor.
- If the estimated reduction in bandwidth due to silence suppression is too aggressive, cell loss could occur. Therefore, the probability of exceeding the MBS will be equated to a scaling factor that describes the associated risk of cell loss. This allows the CLR of the selected QoS category to be maintained.
- The Cell Delay Variation (CDV) incurred by a number of connections will directly relate to the buffer size, as shown in ITU-T I.356 [65]. The contribution made by queuing delay to the maintenance of the target CTD and CDV must also be considered as a second main objective.

Chapter

8

Traffic Modeling at the Talkspurt Level

Both the duration of talkspurt and silence periods are assumed to be exponentially distributed. A unique feature of the exponential distribution is its "memoryless" (Markovian) property. It is important to show how this property exists in the world of telephony and to demonstrate that the exponential distribution is a valid model to use in the characterization of the behavior of a voice source.

This Markovian property also influences the estimated size of the *Peak Cell Rate* (PCR) and *Maximum Burst Size* (MBS) traffic descriptors. A burst of cells from a number of sources that are concurrently in a talkspurt defines the worst-case bandwidth requirement described by the setting of the PCR traffic descriptor.

It is necessary to evaluate the probable lifetime of a talkspurt in order to establish the length (in cells) of the overlap of talkspurts from a number of sources and the estimated worst-case size of a burst of talkspurts to be reflected in the setting of the MBS traffic descriptor. Underestimation of these traffic descriptors will have an effect on the target *Cell Loss Ratio* (CLR). The aim is to identify the optimum number of sources to provide bandwidth for in order to stay within a targeted CLR.

8.1 Lifetime of a Talkspurt

Consider the length of a talkspurt as a continuous random variable X, where X represents the lifetime of the talkspurt. The probability that the talkspurt will live for at least $s + t$ seconds, given that it has already survived s seconds, is the same as the initial probability that it will live for at least t seconds.

In other words, if the talkspurt is alive at time s, then the distribution of its remaining life is the original lifetime distribution —that is, the distribution of remaining life for an s-seconds-old talkspurt is the same as for a new talkspurt and the rate (λ) of transitioning from talkspurt to silence does not change, despite the point in time during the talkspurt at which this probability is evaluated.

Considering X as the lifetime of the talkspurt and supposing that the talkspurt has survived for some t milliseconds, the probability that it will not survive for an additional time (T) is as follows:

That is, consider P{X∈ (t, t + T), |X > t}.

$$P\{X \in (t, t+T), |X > t\} = \frac{P\{X \in (t, t+T), X > t\}}{P\{X > t\}} \tag{8-1}$$

$$= \frac{P\{X \in (t, t+T)\}}{P\{X > t\}}$$

$$= \frac{f(t)T}{\overline{F}(t)} = \lambda(t)T$$

That is, $\lambda(t)$ represents the probability intensity that a t-seconds-old talkspurt will end. Hence, $\lambda(t)$ should be a constant. This is confirmed because

$$\lambda(t) = \frac{\lambda e^{-\lambda t}}{e^{-\lambda t}} = \lambda \tag{8-2}$$

8.2 Markovian Methods of Modeling the Buffer Queue

A queue, or a population, can be modeled using the simplest of stochastic processes: a Markov chain.

8.2.1 Continuous Time Markov Chain

The process of a call transitioning from a talkspurt state to a silence state is a continuous time Markov chain. A continuous time Markov chain is a stochastic process that moves from state $n-1$ to state n in accordance with a discrete time Markov chain in such a way that the

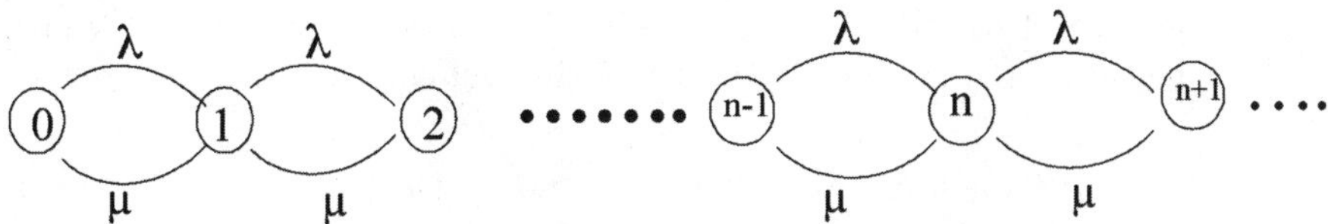

Figure 8-1 Markov chain representation

amount of time spent in each state, before proceeding to the next state, is exponentially distributed.

For example, as illustrated in Figure 8-1, a telephone call will move from a state of talk to a state of silence with probability λ. The call can move from a state of silence to a state of talk with probability μ. Hence, the talk state endures for a random time, with exponential distribution of parameter λ, and then jumps to a silence state with an exponential distribution of parameter μ.

When the Markovian (memoryless) property is applied to a continuous time Markov chain, the future state at time $s + t$, given the present state at s and all past states, depends only on the present state and is independent of the past.

Hence, when a telephone call represented by a continuous time Markov chain enters a state of talk at some time (for example, when $t = 0$ and supposing that the call does not leave the state of talk during the next T milliseconds [ms]), the probability that the talkspurt will not leave the state of talk during the next following T ms is the probability that the talkspurt remains in a talk state during the interval t, $t + T$. This is just the unconditional probability that the call stays in a talk state for at least T ms. This theory can then be extended further to the cell level.

8.2.2 Probability of the Last Cell

The geometric distribution provides the probability that any cell slot belonging to a talkspurt may be the last cell slot before a silence state cell slot is received, that is, the beginning of silence.

As previously shown, the probability that a cell will be the last cell of a talkspurt does not depend on how long the talkspurt has continued due to the memoryless property of the exponential distribution. In other words, the probability that a talkspurt will end is independent of the number of cells that have already been sent in the state of talk.

Example Let τ_k be the probability that the kth cell from the beginning of a talkspurt is the last cell of the current talkspurt and F equals the duration of the cell in seconds. Because a talkspurt is governed by an exponential distribution with parameter λ ($1/\lambda$ being 1.34 sec), the probability of the last cell is

$$\begin{aligned} \tau_k &= P[talkspurt < kF \mid talkspurt \geq (k-1)F)] \qquad (8\text{-}3) \\ &= P[(k-1)\, F \leq talkspurt < kF] \,/P\, [talkspurt \geq (k-1)F] \\ &= 1 - e^{-\lambda F} \\ &= \tau \end{aligned}$$

Notice that the transition probability τ is a frame-independent term to form a geometric process, a phenomenon intuitively clear from the memoryless property of an exponential distribution.

A frame-independent probability σ of a silence period can similarly be derived to yield $\sigma = 1 - e^{-\mu F}$, where μ denotes the parameter of the exponential distribution for the silence period ($1/\mu$ = 1.67 sec).

8.2.3 Birth and Death Process

A continuous time Markov chain is called a *birth and death process* when transitions from state n can only go to state $n-1$ or $n + 1$. There are birth and death processes in several aspects of a voice network:

- There is a birth and death process in the setup and termination of telephone calls, that is, the population of calls that are ongoing. For example, the size of the population of calls in talkspurt increases by 1 when a birth occurs as a channel enters the state of talk and decreases by 1 when a death occurs—that is, when a channel leaves the state of talk and goes into the silence state.
- There is a birth and death process in the transitioning from talkspurt to silence (or vice versa), that is, the population of ongoing calls that are simultaneously in talkspurt (or silence).
- There is a birth and death process in the processing of the sampled voice. Each voice sample upon arrival goes directly out of the egress buffer of the ingress linecard if bandwidth is available; otherwise, the encapsulated voice sample joins a queue in the buffer. When the server process at the egress buffer finishes serving a voice sample, the voice sample leaves the system and the next voice sample enters the server process.

8.2.4 A Pure Birth or Death Process

A Poisson arrival process is an example of a pure birth process. In the telephone network, each source acts independently and produces, or "gives birth," to calls at an exponential rate. Also, each call gives birth in a sense to talkspurts and silences at an exponential rate.

The arrival of talkspurts and silences can be considered as Poisson arrival processes in order to understand the population of calls that could be simultaneously in a talkspurt or in a silence state.

The population of channels aggregated into a Virtual Connection and simultaneously in a talkspurt will provide a worst-case bandwidth requirement. If *Call Admission Control* (CAC) procedures were applied in order to estimate the effective bandwidth, the worst-case bandwidth requirement would be described by the PCR traffic descriptor. The length of a burst at PCR is described by the MBS traffic descriptor in cells.

During a state where a number of sources are overlapping in a talkspurt, a number of calls will transition into the next state of silence before the expected end of the talkspurt—that is, a number of sources will continue in a talkspurt at a time t, and a number of sources will not survive and transition into silence.

The probability that a number of calls will transition into silence during the window of 1.34 sec, which is estimated to be the average-length talkspurt, determines the size of the MBS that should be provisioned for.

8.3 Estimating the Number of Calls Concurrently in a Talkspurt

Consider that there are N voice sources in the connection and suppose that the average amount of time a connection spends in a talkspurt is exponentially distributed with rate $\lambda = 1/1.34$ sec.[1] The state ni, where there are i sources already in a talkspurt, will be modeled as a pure death process—that is, a source dies when it leaves the talkspurt. For example, consider a case where $j = 20$ connections begin in a talkspurt (thus, the following statements depend on the precondition that every source is initially in the talkspurt state).

Pij is the probability that exactly i connections that are currently in the talkspurt state will continue in the talkspurt state at

[1] Paul T. Brady, "A Statistical Analysis of On-Off Patterns in 16 Conversations," *Bell System Technical Journal* (January 1968) [31].

a time t later and that exactly $(j-i)$ connections will end—that is, they will die and leave the talkspurt state at time t later.

It follows that if the population of calls in talkspurts at time t starts with i individuals, the number of calls continuing in talkspurt at time $t + T$ will be the sum of i independent and identically distributed geometric random variables. This will therefore have a negative binomial distribution. This probability, *Pij (T)*, where an exact number of sources (i) continue in a talkspurt and that the remainder ($j - i$) end, is given by the following equation:

$$\mathrm{P}ij(T) = \begin{bmatrix} j-1 \\ i-1 \end{bmatrix} \mathrm{e}^{-\lambda \mathbf{T} i}[(1 - \mathrm{e}^{-\lambda \mathbf{T}})^{(j-i)}], j \geq i \geq 1 \qquad (8\text{-}4)$$

Example The probability that at the start $j = 20$ calls are in a talkspurt and that exactly 9 will continue and 11 will end given that the average length of the talkspurt is 1.34 sec (thus $\lambda = 0.74627$), is as follows:

$$Pij(t) = \begin{bmatrix} 20-1 \\ 9-1 \end{bmatrix} (e^{-0.74627\ 9T})[(1 - e^{-0.74627\ T})^{11}]$$

The memoryless behavior of the Markov chain enables you to look forward or backward without taking account of the duration of the state up to the current point. It is assumed that a number of sources were overlapping initially in the talkspurt state, but that by the end of a certain time $t + T$, a number of sources had transitioned into the next state of silence. The negative binomial distribution, as shown in Equation 8-4, is used to evaluate that risk. Equation 8-4 forms a finite geometric series which, for large N, could be approximated by the normal distribution.

8.4 Effective Bandwidth

Effective bandwidth (sometimes called virtual bandwidth) represents the bandwidth a particular *Virtual Channel Connection* (VCC) or *Virtual Path Connection* (VPC) requires in order to achieve the required *Quality of Service* (QoS) over the life of a connection.

When the connection's traffic source is bursty, it is not necessary to reserve bandwidth continuously at the PCR. However, the average cell rate, described as the *Sustainable Cell Rate* (SCR), is considered to be too optimistic as a descriptor of a long-term bandwidth requirement. The effective bandwidth for a connection is generally less than the PCR and more than the SCR; however, it will tend to the SCR. The extent to which it will tend to the SCR depends on the size of a burst of cells described by the MBS parameter.

8.5 A Method for Dimensioning the MBS

If the risk of a certain number of sources continuing beyond the expected talkspurt length is acceptably low, then the MBS can be dimensioned according to the number of sources, which would expire as expected at the end of the average-sized talkspurt.

For example, if $j = 20$ sources are transmitting, and it is established that the sum of the probabilities of S or more sources continuing to talk for more than T seconds is less than or equal to the CLR target, as illustrated in Equation 8-5, then the MBS and PCR traffic descriptors can be sized to accommodate $(S - 1)$ sources in a talkspurt at the same time. The geometric series providing the probability that S or more sources continue in a talkspurt for a selected duration T is shown in the following equation:

$$\sum_{i=S}^{j} Pij(T) = \sum_{i=S}^{j} \begin{bmatrix} j-1 \\ i-1 \end{bmatrix} e^{\lambda Ti}[(1 - e^{-\lambda T})^{(j-i)}] \leq \text{CLR target} \qquad (8\text{-}5)$$

$$j \geq i \geq 1$$

8.6 Setting the PCR, SCR, and MBS Traffic Descriptors

According to a selected risk factor, the setting of the PCR and MBS traffic descriptors can be based on the bandwidth required for a number of voice channels estimated to be concurrently in a talkspurt but dropping out of the state of talk at or before the end of the average-length talkspurt.

If the average length of a talkspurt is exceeded and the burst is longer than expected, then the target CLR attached to the QoS class will not be maintained.

This risk can be described as the risk that the estimated burst (MBS) will be exceeded. The risk can be quantified. As the traffic descriptors are sized to accommodate a number of sources increasing toward the original total number of contributing sources, the risk reduces. This allows for the choice of an acceptable risk of cell loss, which can be chosen in order to maintain the CLR of the selected QoS class.

For example, a user of a high QoS class with a target CLR of 1.0 e-9 will need to provision for a higher number of sources than a user of a lower QoS class with a target CLR of 1.0 e-7.

The evaluation of the probability that a number of sources exceed the average talkspurt length allows for a scaling factor to be established. The scaling factor provides a guide to how many sources should be allowed when calculating the PCR and MBS setting in order to meet a target CLR.

The sizing of the traffic descriptors would normally be based on the average-length talkspurt. Alternatively, the traffic descriptors could be set more pessimistically based on a longer-length talkspurt, for example, 2 sec. The curve in Figure 8-2 is derived from the implementation of the geometric series shown in Equation 8-2 and illustrates how the probability that *all* sources will continue talking beyond the average-length talkspurt (1.34 sec) becomes increasingly smaller as the number of sources is increased up to 30.

Figure 8-3 illustrates how the probability that a number of sources will continue in a talkspurt varies according to the point in time during the talkspurt at which they are evaluated.

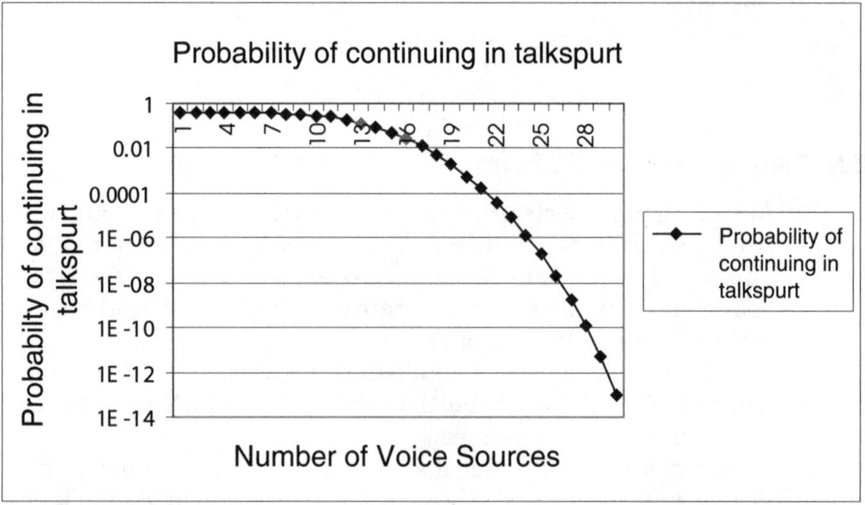

Figure 8-2 Probability that all sources continue in a talkspurt as the number of sources is increased up to 30 (talkspurt = average length)

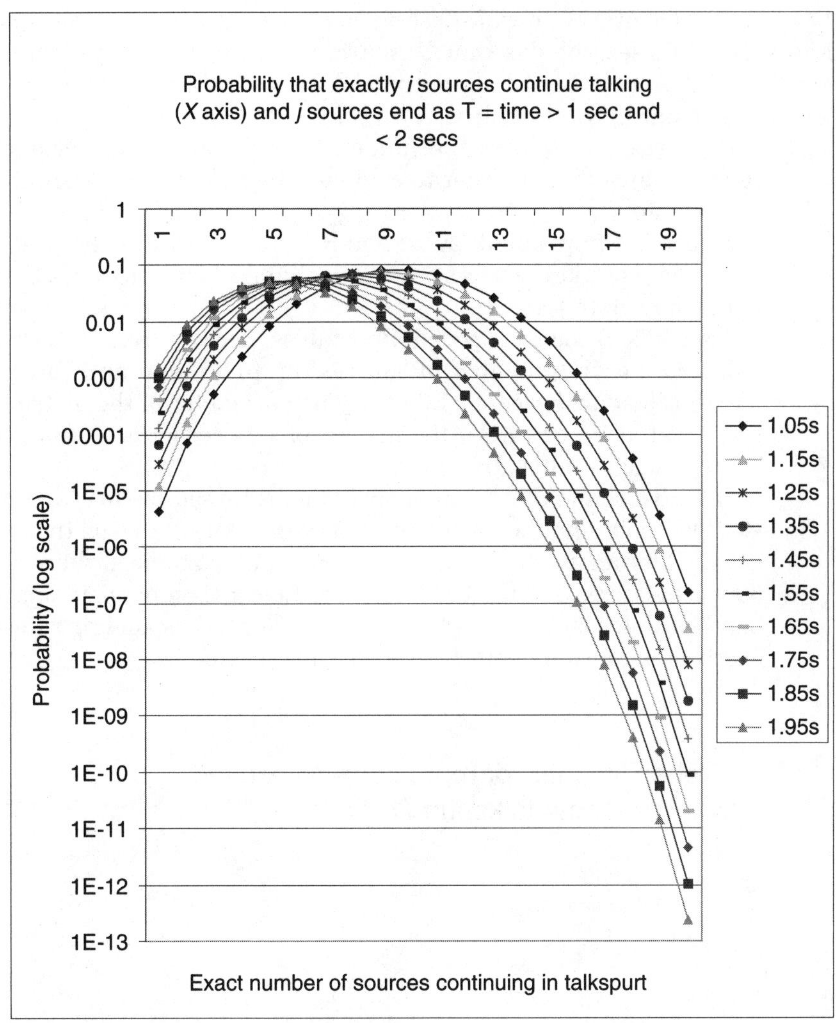

Figure 8-3 Probability that exactly i sources (X axis) out of 20 continue talking for the range $1.0 \leq$ time ≤ 2.0 sec

Given that a number of sources are in a talkspurt at time T, each curve on the graph illustrates the probability, at a time T between 1 and 2 sec (in increments of 0.1 sec), that exactly i sources will continue talking and $(j - i)$ expire.

Figure 8-3 shows how selecting the time at which to evaluate the probability of talk continuing determines the curve that will apply. The graph shows how the risk that a specific number sources continue to talk declines as time extends. For example, after 1.05 sec, the risk that 20 sources continue to talk is circa 1.0 e-7, whereas after 1.95 sec, the risk that 20 sources continue to talk is much lower at circa 1.0 e-13.

It is necessary to take the sum of the probabilities for i or more sources continuing in order to evaluate the risk of sizing the MBS inadequately.

Figure 8-4 implements Equation 8-5 and illustrates the probability that i or more sources continue talking when time $= \tau$ is evaluated at exactly 1.34 sec.

Appendix B contains probability tables, derived from Equation 8-5, for a varying number of sources (up to 30) and an average-length talkspurt based on 1.34 sec. The parameter of the distribution is set at $\lambda = 0.74627$ (the inverse of 1.34 sec for an estimated average-length talkspurt).

Table 8-1 shows the recommended number of sources to provision bandwidth for and the advised rate of overbooking based on the target CLR for an average-length talkspurt. For example, if there are 30 voice sources and the CLR target is 1.0 e-7 then provisioning of bandwidth for 25 sources will take place. The corresponding rate of overbooking recommended is shown in brackets.

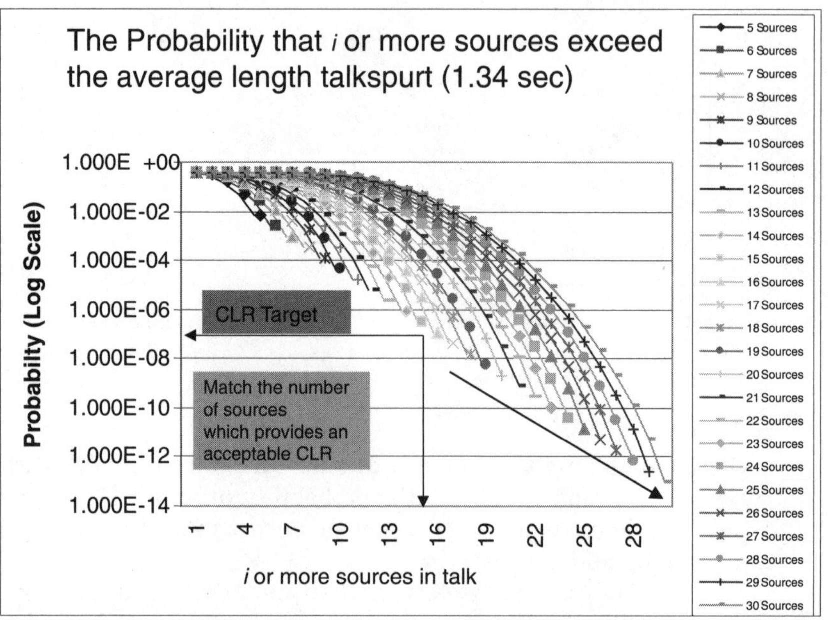

Figure 8-4 The probability that i or more sources (X axis) exceed an average-length talkspurt of 1.34 sec for the range $5 \leq$ number of sources ≤ 30

TABLE 8-1 Number of Sources to Be Provisioned to Achieve the Target CLR when TS = 1.34 sec (Advised Rate of Overbooking)

Number of Voice Sources	Cell Loss Ratio 1.0 e-4	Cell Loss Ratio 1.0 e-5	Cell Loss Ratio 1.0 e-6	Cell Loss Ratio 1.0 e-7	Cell Loss Ratio 1.0 e-8	Cell Loss Ratio 1.0 e-9
10	9 (110%)					
11	10 (109%)					
12	11 (108%)	11 (108%)				
13	11 (115%)	12 (108%)				
14	12 (114%)	13 (107%)	13 (107%)			
15	12 (120%)	13 (113%)	14 (107%)			
16	13 (119%)	14 (113%)	15 (106%)			
17	14 (118%)	15 (112%)	16 (106%)	16 (106%)		
18	14 (122%)	15 (117%)	16 (111%)	17 (106%)		
19	15 (121%)	16 (116%)	17 (111%)	18 (105%)	18 (105%)	
20	15 (125%)	17 (115%)	18 (110%)	18 (110%)	19 (105%)	
21	16 (124%)	17 (119%)	18 (114%)	19 (110%)	20 (105%)	20 (105%)
22	17 (123%)	18 (118%)	19 (114%)	20 (109%)	21 (105%)	21 (105%)
23	17 (126%)	18 (122%)	19 (117%)	20 (113%)	21 (109%)	22 (104%)
24	18 (125%)	19 (121%)	20 (117%)	21 (113%)	22 (108%)	23 (104%)
25	18 (128%)	20 (120%)	21 (116%)	22 (112%)	23 (108%)	23 (108%)
26	19 (127%)	20 (123%)	21 (119%)	22 (115%)	23 (112%)	24 (108%)
27	19 (130%)	21 (122%)	22 (119%)	23 (115%)	24 (111%)	24 (111%)
28	20 (129%)	21 (125%)	22 (121%)	24 (114%)	25 (111%)	25 (111%)
29	20 (131%)	22 (124%)	23 (121%)	24 (117%)	25 (114%)	26 (110%)
30	21 (130%)	22 (127%)	24 (120%)	25 (117%)	26 (113%)	27 (110%)

Conclusions

- Each talkspurt represents a burst of activity when cells are transmitted at a constant rate. The binomial model arises naturally when a number of sources are multiplexed onto the same connection.
- By the memoryless property, the distribution of remaining life for a *t*-seconds-old talkspurt is the same as for a new talkspurt. In other words, $\lambda(t)$ is a constant representing the probability intensity that a new talkspurt or a continuing talkspurt will end in *t* seconds time.
- A worst-case bandwidth requirement occurs when a number of sources are concurrently in a talkspurt—this is equivalent to setting a PCR traffic descriptor. The number of sources under consideration can be the total number of sources contributing to the same link, which is the very worst case, or a reduced number of sources taking into account the probability that a reduced number of sources will be in a talkspurt at the same time.

- When the worst-case bandwidth setting (PCR) is determined by including all contributing sources, the bandwidth requirements may be set too high because only the potential for talkspurts continuing beyond a mean estimated length is taken into account—that is, the worst-case bandwidth requirement is based on the assumption that every single voice channel is concurrently in a talkspurt.
- Assuming that every source could be in a talkspurt at the same time may be reasonable for a small number of sources, but this becomes less reasonable when the number of sources is large.
- However, when the worst-case bandwidth requirement is based on the average number of talkspurts expected to arrive, the number of sources provisioned for is likely to be closer to a Sustained Cell Rate (SCR), or the average rate.
- The question arises as to how much the worst-case bandwidth setting should be reduced. In other words, should this be the average number of sources expected to be in a talkspurt concurrently, or should it be above the average to a degree (which is based on a risk factor)?
- It is possible to evaluate the probability that a selected number of voice sources concurrently in a talkspurt will exceed the average-length talkspurt and will continue in talk rather than transition into silence. This provides a quantifiable risk factor. When this risk is tolerably low, the Maximum Burst Size (MBS) is assessed at the number of voice sources that are estimated to end the talkspurt at the expected time.
- The mathematical approach described in this book enables the separation and quantification of the two main risks—the risk of assessing the probability of a number of talkspurts arriving simultaneously and the risk of assessing the average duration of a talkspurt. Both of these variables are seen to be of exponential duration.
- Given that PCR, SCR, and MBS traffic descriptors have been defined, a CAC algorithm would provide an effective bandwidth that would be less pessimistic than an estimate based on the worst-case bandwidth requirement (PCR) and would provide a guide to how many sources to provision bandwidth for while maintaining the CLR commitment of the QoS class.

Chapter

9

Application of a Call Admission Control (CAC) Algorithm

ATM layer *Quality of Service* (QoS) is a long-term commitment. ITU-T Recommendation I.356 [64] specifies QoS objectives for an end-to-end connection and apportionment rules establishing QoS objectives for each standardized connection portion.

9.1 Call Admission Control (CAC)

A *Call Admission Control* (CAC) process determines the effective bandwidth required for one connection or several multiplexed connections. Connection Admission Control is defined by ITU-T I.371 "Traffic Control and Congestion Control in B-ISDN" [65].

The ATM Forum does not make specific recommendations on which algorithm vendors should use to implement the CAC process. Anick, Mitra, and Sondhi [24] published an analysis for the statistical multiplexing of N sources of a single type onto a communication channel.

This model is called the *Uniform Arrival and Service Model* (UAS) in Daigle and Langford [27]. Gibbens and Hunt applied the uniform arrival and service model in [78], showing that it is possible to assign an effective bandwidth to each source, dependent not only on its mean bandwidth, but also on its burstiness and based on a target cell loss probability.

Several other papers discuss the simple effective-bandwidth approach; for example, see Hui [44], Guerin et al. [38], Kelly [45],

Sohraby [50], [51], Chang [32], Whitt [53], Elwalid and Mitra [36], Kesidis et al. [42], Glynn and Whitt [37], and Courcoubetis et al. [34].

The calculation of effective bandwidth requires inputs of *Peak Cell Rate* (PCR), *Sustainable Cell Rate* (SCR), and *Maximum Burst Size* (MBS). It is proposed in this book that the determination of the PCR, SCR, and MBS traffic descriptors should be based on the mathematical modeling of the behavior of the multiplexed on/off voice sources as described in Chapter 8.

This modeling should be part of the process that seeks to provide a *Cell Loss Ratio* (CLR) guarantee. There is a risk that the estimated traffic descriptors will be exceeded, and this risk is equated to the CLR target.

The CLR targets for each of the *Variable Bit Rate* (VBR) classes are shown in Table 6-1. The Rt-VBR classes have a target *Cell Delay Variation* (CDV) to meet in addition to the CLR constraint.

9.2 Scaling the PCR and MBS Traffic Descriptors

The estimate of the PCR traffic descriptor reflects the worst-case bandwidth requirements and can be based on all sources in a talkspurt or the most likely number of sources in a talkspurt.

However, as the number of sources under consideration is increased, the likelihood that they will all be in a talkspurt concurrently becomes increasingly low, and the PCR traffic descriptor could, in theory, be sized according to a reduced number of sources entering the state of talkspurt concurrently. Thus, the PCR setting may be sized in two ways:

1. As the bandwidth equivalent to the situation where *all* of the sources contributing to the same AAL interface are in a talkspurt concurrently
2. As the bandwidth equivalent to the number of sources *likely* to continue throughout a talkspurt together based on the mathematical probability method described in Chapter 8.

In the second option, when the risk that a number of sources will continue talking for longer than the estimated talkspurt (1.34 sec) is acceptably low, the balance of sources remaining becomes the basis for sizing the MBS and PCR traffic descriptors. The MBS traffic descriptor is defined as the size of a burst of cells.

The size of a single talkspurt in cells depends on the estimated length of the talkspurt in seconds, the framing overhead (which depends on the type of adaptation layer in use), and the effect of combining a number of coder samples in one single adaptation layer frame (which in turn depends on the packing factor if AAL5 is used or the timer setting if AAL2 is used).

In all cases, the SCR is set to 45 percent of the collective bandwidth requirements of all the sources contributing to the interface. Table 9-1 shows the number of cells in a talkspurt when AAL5 is used with a variety of different packing factors and the equivalent bandwidth for setting the PCR traffic descriptor. The average talkspurt size is estimated at 1.34 and 2 sec. The formulas used in Table 9-1 can be found in Appendix H of this book.

Table 9-2 shows the number of cells in a talkspurt when AAL2 is used with a varying number of channels contributing to a virtual connection and the equivalent bandwidth for setting the PCR traffic descriptor. The average talkspurt size is estimated at 1.34 and 2 sec. The formulas used in Table 9-2 can be found in Appendix H.

9.3 Application of CAC

A CAC algorithm will be used to provide a guide to the following question:

> Given a number of sources estimated as being in the state of talkspurt concurrently and given a related size of MBS according to this number of sources, the size of the coder unit, packing factor or timer setting, and given that all of the sources contributing to the connection will experience an average gain due to silence removal of circa 55 percent (Brady [31]), described by the SCR parameter, what is the effective bandwidth that should be provided for, in those circumstances?

9.4 Comparison of Effective Bandwidth Versus CLR Target and Queue Length

Using the mathematical probability method for sizing a reduction in bandwidth due to silence removal, as described in Chapter 8 of this book, it is established that for 30 voice sources and a CLR target of 1.0 e-6 (equivalent to the QoS class of Nrt-VBR Class B), the PCR and MBS traffic descriptors should be sized for a reduced number of 24 sources.

TABLE 9-1 An Average Talkspurt Size (1.34 and 2 sec) in Cells for 10 Coder Types Using AAL5

Coder Type	Output Rate of Coder (bps)	Coder Unit (ms)	Packing Factor Range	Packing Factor Selected	Number of ATM cells for each AAL5 frame	Bandwidth Required Incl. Overheads (Kbps)	Number of Cells in a Talkspurt of 1.34 sec	Number of Cells in a Talkspurt of 2.0 sec
Clear channel 64 Kbps	64,000	5	1 to 4	2	2	84.80	269	401
G.711 μ-law 64 Kbps	64,000	5	1 to 4	2	2	84.80	269	401
G.711 A-law 64 Kbps	64,000	5	1 to 4	1	2	169.60	537	801
G.723.1 5.3 Kbps	5,300	30	1 to 6	3	2	9.42	31	45
G.723.1 6.3 Kbps	6,300	30	1 to 7	7	4	8.08	27	39
G.726 16 Kbps	16,000	5	1 to 18	4	2	42.40	135	201
G.726 24 Kbps	24,000	5	1 to 15	2	1	42.40	135	201
G.726 32 Kbps	32,000	5	1 to 9	3	2	56.53	180	268
G.726 40 Kbps	40,000	5	1 to 7	1	1	84.80	269	401
G.729A 8 Kbps	8,000	10	1 to 13	10	3	12.72	41	61

TABLE 9-2 An Average Talkspurt Size (1.34 and 2 sec) in Cells for 10 Coder Types Using AAL2

Coder Type	Output Rate of Coder (bps)	Coder Unit (ms)	Number of Channels Contributing to AAL2 Interface	Choose Max. Timer Setting (ms)	Recomm. Timer Setting for Full Cell Fill	Bandwidth Required for all Channels (Kbps)	Average Bandwidth per Single Channel (Kbps)	Number of Cells in a Talkspurt of 1.34 sec	Number of Cells in a Talkspurt of 2 sec
Clear channel 64 Kbps	64,000	5	2	3	2.50	155.17	77.58	515	768
G.711μ-law 64 Kbps	64,000	5	26	1	0.20	2,017.16	77.58	6,679	9,968
G.711 A-law 64 Kbps	64,000	5	26	1	0.20	2,017.16	77.58	6,679	9,968
G.723.1 5.3 Kbps	5,300	30	26	3	2.31	178.85	6.88	628	936
G.723.1 6.3 Kbps	6,300	30	26	2	1.16	208.17	8.01	718	1,071
G.726 16 Kbps	16,000	5	1	1	15.00	84.80	84.80	91	135
G.726 24 Kbps	24,000	5	5	1	2.00	162.38	32.48	588	876
G.726 32 Kbps	32,000	5	12	1	0.84	497.97	41.50	1,743	2,601
G.726 40 Kbps	40,000	5	20	1	0.25	1,010.38	50.52	3,463	5,168
G.729A 8 Kbps	8,000	10	12	2	2.50	140.73	11.73	537	801

The following traffic descriptor settings would apply to 24 sources using a G.726 32-Kbps coder:

- PCR set to 2.0352 Mbps (24×84.8 Kbps)
- MBS sized at 6,456 cells (269 cells in each talkspurt×24 sources)
- SCR set to 1.1448 Mbps (30×84.8 Kbps×0.45)

Figure 9-1 shows the effective bandwidth recommended according to the CAC algorithm. It is spurious accuracy to size a PCR

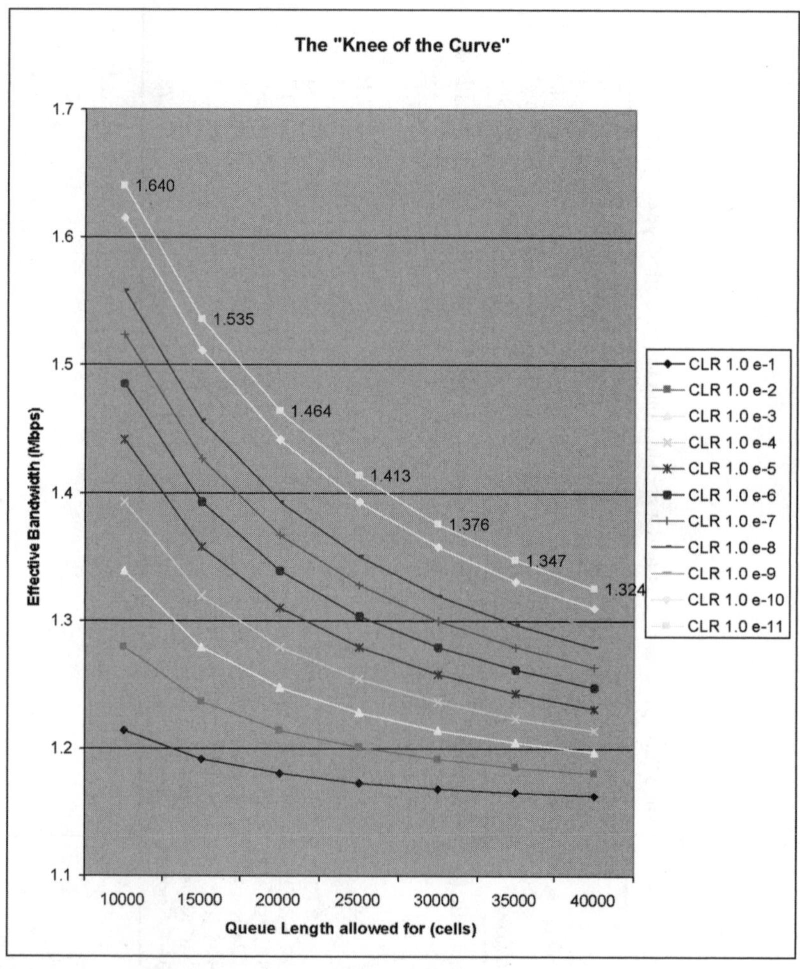

Figure 9-1 Effective bandwidth for a range of CLR targets and queue lengths PCR = 2.0352 Mbps (AAL5) (24 sources), MBS = 6,456 cells (24 sources), and SCR = 1.1448 Mbps (30 sources)

and MBS according to a CLR target and then to weaken that CLR target in further calculations. Hence, only the effective-bandwidth results for the same CLR target or a higher (more stringent) CLR target are of interest.

Figure 9-1 also shows how effective bandwidth has to be provisioned in increasing amounts to meet the more challenging CLR targets for the shorter queue lengths and how effective bandwidth results for each CLR target are converging for the longer queue lengths.

After a point, at the knee of the curve, the opportunity to trade a lower-effective bandwidth for a higher queue length declines, as all the CLR curves converge. The *knee* of the curve is the area where the optimal *trade-off* for queue length and effective bandwidth can be found. However, delay and CDV constraints could mean that a wide range of queue lengths are not necessarily viable.

9.5 Calculating Effective Bandwidth for an Nrt-VBR Service

The results illustrated in Table 9-3 show the effective bandwidth provisioned by the CAC algorithm when the queue in the buffer is allowed to extend to 5,000 cells, which is appropriate for an Nrt-VBR Class A service queue being served by an STM-1 trunk with a delay due to the queuing and shaping of circa 19 ms.[1]

Detailed Guide to Results in Table 9-3 Table 9-3 illustrates the results when the CAC algorithm is applied using two settings for the PCR and MBS traffic descriptors:

- PCR and MBS are set for *all* sources in a talkspurt.
- PCR and MBS are set for a reduced number of sources in a talkspurt where the reduction has been worked out using the mathematical probability method described in this book.

Column 1 Shows the effective bandwidth (in an equivalent number of sources) output by the CAC when PCR is set according to *all* sources contributing to the interface and when SCR is set according to 45 percent of the bandwidth required by PCR.[2] MBS is set at a rate of all sources continuing throughout a talkspurt together as

---

[1] This delay result is based on the queue being emptied at PCR during a burst and SCR at other times. See Table 9-4.

[2] Forth-five percent is based on Brady's [31] research results.

TABLE 9-3 Effective Bandwidth for a Nrt-VBR Service and a Queue Length of 5000 Cells

Comparison of effective bandwidth provisioned for all sources, compared to effective bandwidth provisioned for a reduced number of sources according to the mathematical probability method for a G.726 coder using AAL5 with a packing factor of 1 where the burst size is equivalent to 269 cells per source, PCR = 84,800 bps, all sources = 30, reduced sources = 22 (CLR 1.0 e-5), 24 (CLR 1.0 e-6), 25 (CLR 1.0 e-7), queue size = 5,000 cells

	(1)	(2)	(3)	(4)	(5)	(6)	(7)
Cell Loss Ratio Target	Effective Bandwidth Output (in sources) Using PCR and MBS of All 30 Sources, Queue Length 5,000 Cells (Nrt-VBR)	Effective Bandwidths % of All Sources	Number of Sources Provisioned Using Mathematical Probability Method to Reduce PCR and MBS	Mathematical Method (in 3) as % of All Sources	Effective Bandwidth Output (in sources) with the Reduced Number of Sources in PCR and MBS as Shown in (3)	(5) as % of (3) sources in MBS/PCR Reduced by Mathematical Probability Method (that is, expected reduction)	(5) as % of All 30 Sources
CLR 1.0 e-5	24	80%	22	73%	18	82%	60%
CLR 1.0 e-6	25	83%	24	80%	20	83%	67%
CLR 1.0 e-7	25	83%	25	83%	21	84%	70%

(coder unit size [cells] according to packing factor/timer setting) × the total number of sources. Example (Row 1): 30 voice sources were aggregated in a connection with a CLR target of 1.0 e-5, and the PCR and MBS traffic descriptors were configured to the bandwidth equivalent for all 30 sources. The SCR traffic descriptor was configured to the bandwidth equivalent to 14 sources (45 percent). The CACed bandwidth requirement output was the equivalent of 18 sources, as shown in Column 2, that is, a reduction to 60 percent of PCR.

Column 2 Shows the equivalent number of sources in the effective bandwidth as a percentage of the original total number of sources when the CAC is applied without adjusting the traffic descriptors, as in Column 1.

Column 3 Shows the number of sources that should be allowed concurrently in a talkspurt according to the probability tables for silence removal (see Appendix B) and in setting the PCR and MBS traffic descriptors.

Column 4 Shows the reduced number of sources shown in Column 3 as a percentage of the original total number of sources.

Column 5 Shows the equivalent number of sources in the effective bandwidth output by the CAC when PCR and MBS are set according to the reduced number of sources likely to continue in a talkspurt derived from the mathematical probability method.

Column 6 Shows the equivalent number of sources in the effective bandwidth shown in Column 5 as a percentage of the number of sources reduced for silence removal, as shown in Column 3.

Column 7 Shows the equivalent number of sources in the effective bandwidth shown in Column 5 as a percentage of the original total number of sources. This reduction therefore includes both the effects of gain to a specific adjustment for silence removal and gain due to statistical multiplexing.

9.6 Analysis of Results for Nrt-VBR

The results in Table 9-3 show that if a pessimistic approach were taken and all 30 contributing sources were included in the PCR/MBS setting, the CAC would provide an effective bandwidth that is equivalent to 24–25 sources, which is equivalent to 80–83 percent of the original bandwidth requirement for 30 sources at PCR. However, if the number of sources in PCR and MBS are first reduced using the mathematical probability method described in this book, a greater reduction to between 60 and 70 percent (CLR 1.0 e-5 to CLR 1.0 e-7) of the original bandwidth can be achieved.

9.7 Maximal-Rate Envelope

A maximal-rate envelope for viable effective bandwidths that offer some statistical gain can be defined within the constraints of

- A range of acceptable delays due to queuing at the egress buffer defined by the user's delay budget
- A range of acceptable CLR targets, at least as stringent as the baseline CLR established by the mathematical probability method
- A range of queue lengths mapped to the range of acceptable CLR targets
- A CDV constraint implied by the use of an Rt-VBR class

Hence, the maximal-rate envelope defines the set of effective bandwidth rates achievable within the previously mentioned constraints and shows optimal settings in terms of statistical gain.

Table 9-4 provides the long-term average delay encountered by the last cell in a queue, according to the egress trunk speed, based on emptying the queue at PCR during an average burst lasting 1.34 sec and emptying the queue at SCR during an average silence lasting 1.67 sec.

TABLE 9-4 Average Delay According to Queue Length and Speed of Egress Trunk[a]

Egress Trunk Type	T1	E1	E3	DS3 PLCP	OC3
Bandwidth (Mbps)	1.544	2.048	34	44.736	155
PCR (cells per second)	3,300	4,400	73,000	96,000	353,207
SCR (cells per second)	1,485	1,980	32,850	43,200	158,943
Average rate (cells per second)	2,293	3,057	50,724	66,706	245,426
Buffer/Queue Length (Cells)	**Average Delay in ms if Emptied Deterministically at PCR During the MBS and SCR at Other Times**				
10	3.668	2.935	0.166	0.138	0.038
20	7.336	5.870	0.332	0.277	0.075
30	11.004	8.805	0.498	0.415	0.113
40	14.672	11.739	0.664	0.554	0.150
50	18.340	14.674	0.831	0.692	0.188
60	22.008	17.609	0.997	0.831	0.226

Egress Trunk Type	T1	E1	E3	DS3 PLCP	OC3
Buffer/Queue Length (Cells)	Average Delay in ms if Emptied Deterministically at PCR During the MBS and SCR at Other Times				
70	25.68	20.54	1.16	0.97	0.26
80	***29.34***	23.48	1.33	1.11	0.30
90	***33.01***	26.41	1.50	1.25	0.34
100	***36.68***	***29.35***	1.66	1.38	0.38
150	***55.02***	***44.02***	2.49	2.08	0.56
200	73.36	***58.70***	3.32	2.77	0.75
300	110.04	88.05	4.98	4.15	1.13
400	146.72	117.39	6.64	5.54	1.50
500	183.40	146.74	8.31	6.92	1.88
600	220.08	176.09	9.97	8.31	2.26
700	256.76	205.44	11.63	9.69	2.63
800	293.44	234.79	13.29	11.07	3.01
900	330.12	264.14	14.95	12.46	3.39
1,000	366.80	293.49	16.61	13.84	3.76
2,000	733.59	586.97	***33.22***	***27.69***	7.52
3,000	1,100.39	880.46	***49.83***	***41.53***	11.29
4,000	1,467.19	1,173.94	66.45	***55.37***	15.05
5,000	1,833.98	1,467.43	83.06	69.21	18.81
6,000	2,200.78	1,760.91	99.67	83.06	22.57
7,000	2,567.58	2,054.40	116.28	96.90	26.34
8,000	2,934.37	2,347.89	132.89	110.74	***30.10***
9,000	3,301.17	2,641.37	149.50	124.58	***33.86***
10,000	3,667.96	2,934.86	166.11	138.43	***37.62***
15,000	5,501.95	4,402.29	249.17	207.64	***56.44***
20,000	7,335.93	5,869.72	332.23	276.85	75.25
30,000	11,003.89	8,804.57	498.34	415.28	112.87
40,000	14,671.86	11,739.43	664.45	553.71	150.50

[a] Viable queue lengths are considered to be where the delay is in the region of 30 to 50 ms. These are bold and italic.

9.8 Tolerable Queue Length Due to Delay

The *tolerable queue length* is determined by the product of the delay budget and the available service rate. A budgeted queuing delay in the range of 30 to 50 ms is a typical requirement for voice.

In Table 9-4, the ceiling of the range of permissible queue lengths on the grounds of delay are shown in bold and italic for each trunk type. Acceptable queue lengths within the delay constraints are approximately 100 cells for a T1 trunk type, 120 cells for an E1, 3,000 for an E3/ DS3 trunk type, and 10,000 for an STM-1 trunk type. However, Rt-VBR traffic classes must also meet a CDV constraint, which will further restrict the available queue lengths.

Table 9-5 takes into account the additional constraints on the queue lengths to meet the one-point CDV requirement for Rt-VBR classes[3] of 250 ms for Rt-VBR Class A and 2,500 ms for Rt-VBR Class B. (For class definitions, refer to Table 6-1). The one-point CDV is calculated at PCR.

For example, to meet the CDV constraint for Rt-VBR Class A, the OC3 CAC buffer is computed to be (250 e-6 sec) × (353,207 cells per sec) = 88.3 ~ 88 cells. To meet the CDV constraint for Rt-VBR Class B, the OC3 CAC buffer is computed to be (2,500 e-6 sec) × (353,207 cells per sec) = 883 cells. There are no CDV constraints for Nrt-VBR classes, and the queue lengths shown are to meet a targeted queuing delay of circa 30 to 50 ms as shown in Table 9-4.

Table 9-6 illustrates the reduced number of sources provisioned by the mathematical probability method for a CLR target when there are 30 sources (refer to Table 8-1).

9.8.1 Effective Bandwidth for an Rt-VBR Class of Service

Figure 9-2 illustrates the effective bandwidth provisioned for 30 sources aggregated into a single connection, when a Rt-VBR Class

TABLE 9-5 Queue Lengths Subject to Delay Tolerances by QoS and Egress Line Rate

Class of Service	T1	E1	E3	DS3 PLCP	OC3
Rt-VBR Class A	***1***	***1***	***20***	***24***	***88***
Rt-VBR Class B	***8***	***11***	***200***	***240***	***883***
Nrt-VBR Class A	100	120	3,000	3,000	10,000
Nrt-VBR Class B	100	120	3,000	3,000	10,000
Nrt-VBR Class C	100	120	3,000	3,000	10,000

TABLE 9-6 Number of Sources (% of All 30 Sources) Reduced by the Mathematical Probability Method for Silence Removal According to a CLR Target

Number of Voice Sources	CLR 1.0 e-4	CLR 1.0 e-5	CLR 1.0 e-6	CLR 1.0 e-7	CLR 1.0 e-8	CLR 1.0 e-9	CLR 1.0 e-10	CLR 1.0 e-11
30	21 (70%)	22 (73%)	24 (80%)	25 (83%)	26 (87%)	27 (90%)	28 (93%)	28 (93%)

[3] This is shown in bold and italic in Table 9-5.

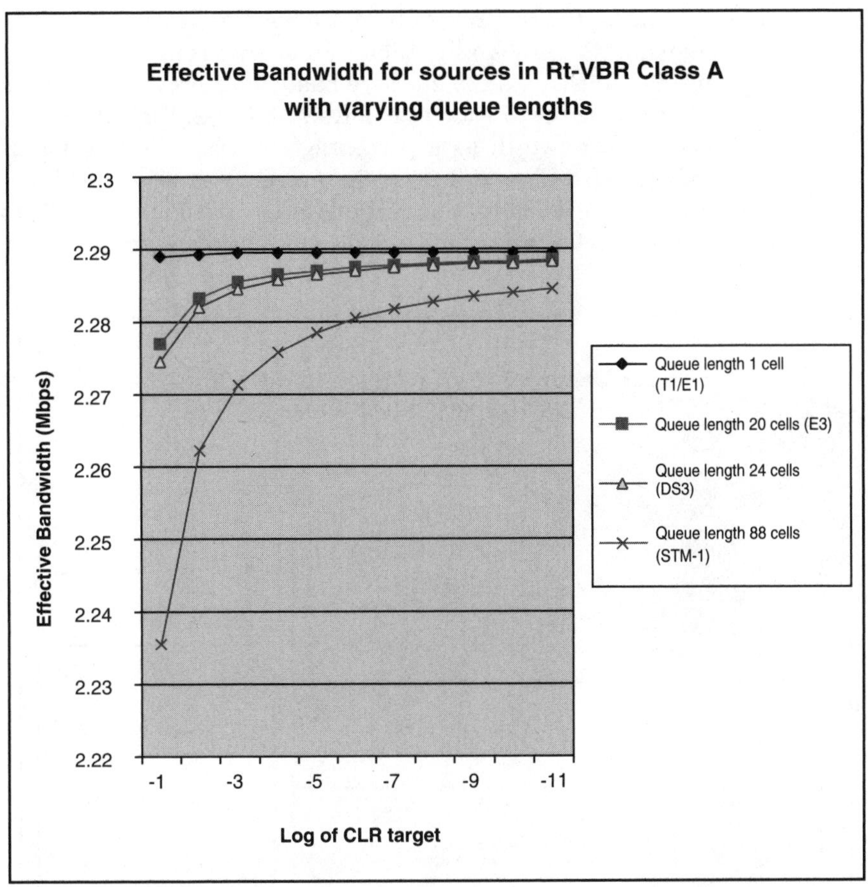

Figure 9-2 Effective bandwidth for 30 sources in Rt-VBR Class A with PCR and MBS reduced to the bandwidth equivalent to 27 sources

A service is selected and the CLR target is 1.0 e-9. The PCR and MBS traffic descriptors are sized for a reduced number of 27 sources (2.29 Mbps at PCR), according to the mathematical probability method described here, and the queue length is constrained to the length appropriate to meet a CDV of 250 ms. SCR is sized at 1.145 Mbps (average gain due to silence removal for 30 sources at 84.8 Kbps = 2.544 × 0.45).

Each curve in Figure 9-2 represents the effective bandwidth that would be provisioned for a different trunk type. Application of the CAC does not allow a significant further reduction in the bandwidth; however, a reduction of 10 percent has already been realized through the provision of 27 sources, rather than 30 sources, using the probability tables for silence removal.

Figure 9-3 illustrates the effective bandwidth provisioned when the CAC is applied to 27 sources in a Rt-VBR Class. Table 9-7 shows the effective bandwidth as a percentage of the PCR rate for 27 sources in Rt-VBR Class A. Table 9-8 shows the corresponding information for Rt-VBR Class B. A CLR target of 1.0 e-9 applies to both these classes and these results are shown in bold and italic.

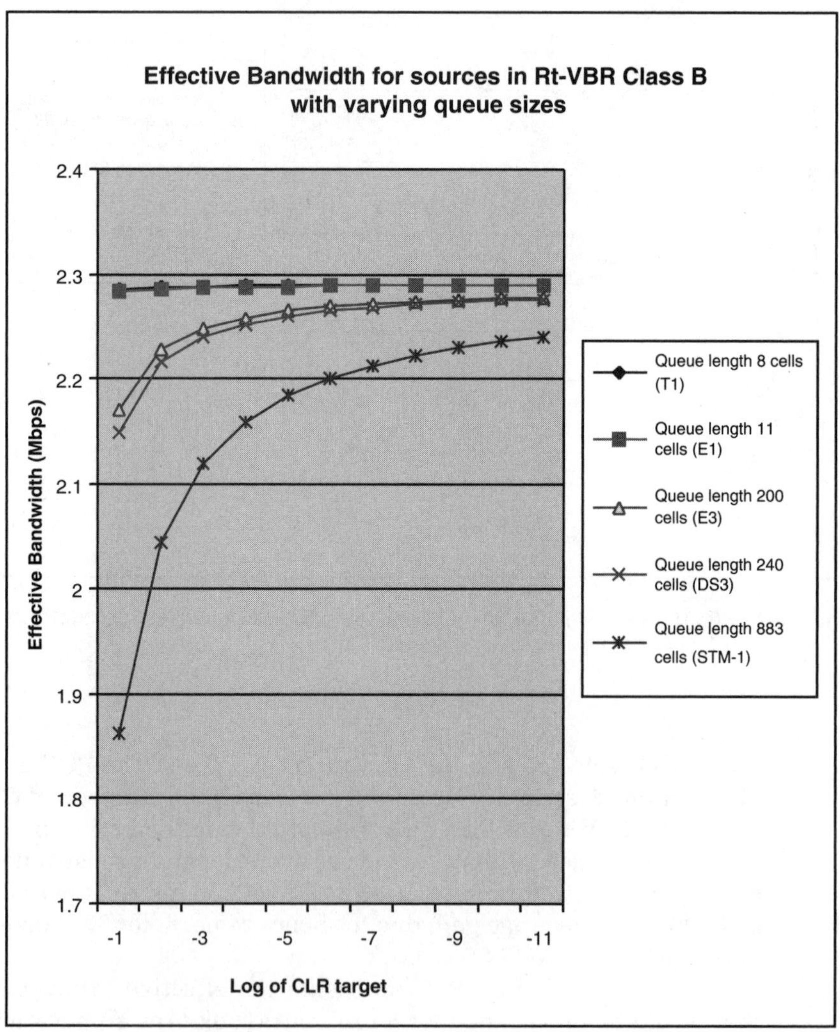

Figure 9-3 Effective bandwidth for 30 sources in Rt-VBR Class B with PCR and MBS reduced to the bandwidth equivalent to 27 sources

TABLE 9-7 Effective Bandwidth as a Percentage of PCR Bandwidth for 27 Sources in an Rt-VBR Class A Service Class

Trunk Type	T1/E1	E3	DS3	STM-1	Trunk Type	T1/E1	E3	DS3	STM-1
Queue/buffer length (cells)/CLR target	1	20	24	88	**Queue/buffer length (cells)/CLR target**	1	20	24	88
CLR 1.0 e-1	100%	99%	99%	98%	**CLR 1.0 e-6**	100%	100%	100%	100%
CLR 1.0 e-2	100%	100%	100%	99%	**CLR 1.0 e-7**	100%	100%	100%	100%
CLR 1.0 e-3	100%	100%	100%	99%	**CLR 1.0 e-8**	100%	100%	100%	100%
CLR 1.0 e-4	100%	100%	100%	99%	***CLR 1.0 e-9***	***100%***	***100%***	***100%***	***100%***
CLR 1.0 e-5	100%	100%	100%	100%	**CLR 1.0 e-10**	100%	100%	100%	100%

TABLE 9-8 Effective Bandwidth as a Percentage of PCR Bandwidth for 27 Sources in an Rt-VBR Class B Service Class

Trunk Type	T1	E1	E3	DS3	STM-1	Trunk Type	T1	E1	E3	DS3	STM-1
Queue/buffer length (cells)/CLR target	8	11	200	240	883	**Queue/buffer length (cells) / CLR target**	8	11	200	240	883
CLR 1.0 e-1	100%	100%	95%	94%	81%	**CLR 1.0 e-6**	100%	100%	99%	99%	96%
CLR 1.0 e-2	100%	100%	97%	97%	89%	**CLR 1.0 e-7**	100%	100%	99%	99%	97%
CLR 1.0 e-3	100%	100%	98%	98%	93%	**CLR 1.0 e-8**	100%	100%	99%	99%	97%
CLR 1.0 e-4	100%	100%	99%	98%	94%	***CLR 1.0 e-9***	***100%***	***100%***	***99%***	***99%***	***97%***
CLR 1.0 e-5	100%	100%	99%	99%	95%	**CLR 1.0 e-10**	100%	100%	99%	99%	98%

The effective bandwidth provisioned for a CLR of 1.0 e-9 makes a slightly further reduction of 1 percent (based on 27 sources) for the longer queue length applicable to an E3/DS3 trunk and a reduction of 3 percent (based on 27 sources) for the queue length applicable to an STM-1 trunk.

9.8.2 Effective Bandwidth for an Nrt-VBR Class of Service

Figure 9-4 illustrates the effective bandwidth provisioned when 30 sources are combined in a connection in Nrt-VBR Class B with a CLR target of 1.0 e-6. The CAC is applied with PCR and MBS configured bandwidth equivalent to 24 sources. Table 9-9 shows the effective bandwidth as a percentage of the PCR rate for 24 sources in Nrt-VBR Class B. Application of the CAC process

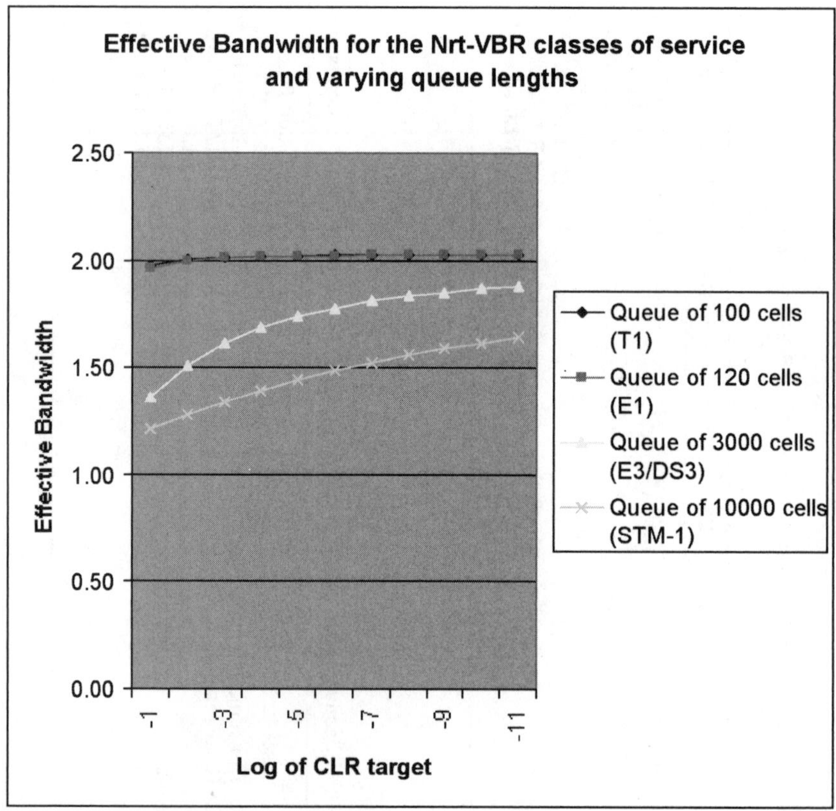

Figure 9-4 Effective bandwidth for 30 sources in Nrt-VBR Class B with PCR and MBS reduced to the bandwidth equivalent to 24 sources

TABLE 9-9 Effective Bandwidth as a Percentage of PCR Bandwidth for Sources in an Nrt-VBR Class B Service Class

Trunk Type	T1	E1	E3/DS3	STM-1	Trunk Type	T1	E1	E3/DS3	STM-1
Queue/buffer length (cells)/CLR target	100	120	3,000	10,000	**Queue/buffer length (cells)/CLR target CCL**	100	120	3,000	10,000
CLR 1.0 e-1	97%	96%	67%	60%	**CLR 1.0 e-6**	99%	99%	87%	73%
CLR 1.0 e-2	98%	98%	74%	63%	***CLR 1.0 e-7***	***100%***	***99%***	***89%***	***75%***
CLR 1.0 e-3	99%	99%	79%	66%	**CLR 1.0 e-8**	100%	100%	90%	77%
CLR 1.0 e-4	99%	99%	83%	68%	**CLR 1.0 e-9**	100%	100%	91%	78%
CLR 1.0 e-5	99%	99%	85%	71%	**CLR 1.0 e-10**	100%	100%	92%	79%

causes no further reduction on the bandwidth on the basis of a queue sized for a T1 trunk type. A reduction in bandwidth of 1 percent becomes possible for the queue appropriate to a T1 or an E1 trunk type. A reduction of approximately 13 percent becomes possible for the queue applicable to an E3/DS3 trunk type, and reductions of approximately 27 percent are applicable when the queue is served by an STM-1 trunk.

9.9 The Importance of Scaling Down the Traffic Descriptors

The Nrt-VBR classes offer the most interesting potential for further statistical gain because the queue lengths are allowed to be longer. Rt-VBR Class A is restricted by the CDV constraint to such small queue lengths that there is little or no further reduction allowed in the bandwidth when a CAC process is applied. The importance of scaling down the traffic descriptors to take account of silence removal before applying the CAC algorithm has been emphasized. Without scaling down, the opportunity to realize statistical gain within the Rt-VBR classes may be strongly restricted by the CDV constraint.

Figure 9-5 illustrates how the mathematical probability method scales down the effective bandwidth. Each curve illustrated is a lower CLR target. The probability that a number of talkspurts will burst the buffer is the same as the CLR target. For example, when there are 30 sources and the MBS is sized for the probability that no more than 22 sources will continue in a talkspurt, then the risk that this MBS is undersized is 1 in 100,000 (CLR 1.0 e-5).

The effective bandwidth is then sized according to this MBS and to achieve a maximum delay dependent on the speed of the egress trunk and the permissible buffer queue size. The risk that this delay will be exceeded is again the same as the CLR target.

For example, as shown in Table 9-10, for a CLR of 1.0 e-5, if the buffer is sized for 1,000 cells, the application of the CAC using a PCR and MBS sized for 22 sources produces an effective bandwidth of 1.75 Mbps (see Row A). This result is 68.8 percent (Row E) of the bandwidth at PCR for all 30 sources. Hence, the application of the mathematical probability method to size a reduction for silence removal, combined with application of the CAC, causes a total reduction in bandwidth, in this case 31.2 percent.

The larger part of the 31.2 percent gain is attributable to the initial reduction in the number of sources, provisioned for in PCR and MBS, from 30 to 22 due to silence removal. Row B shows that applying the CAC to 22 sources results in an effective bandwidth, which is 93.8 percent of the PCR for 22 sources for a buffer of 1,000

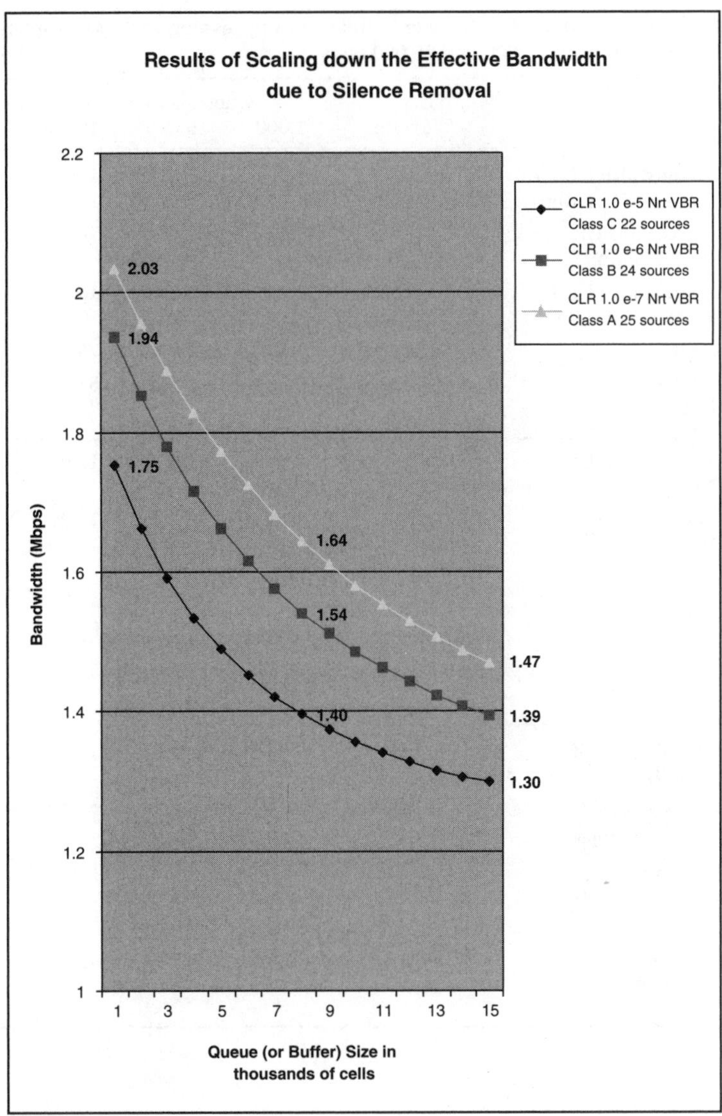

Figure 9-5 Effect of scaling down the effective bandwidth due to silence removal for Nrt-VBR Classes A through C

cells—that is, the CAC process has only reduced PCR by 6.2 percent.

If one preferred to provision for the possibility that all sources are concurrently in a talkspurt, then the effective bandwidth provisioned by the CAC when there are 30 sources and when the

TABLE 9-10 Bandwidth Reduction Due to Silence Suppression and CAC Application for Buffers Sized Between 1,000 and 15,000 cells

CLR Target	Key to Results	Queue/Buffer Size in Cells 1,000	5,000	10,000	15,000
Effective bandwidth CLR 1.0 e-5 (Nrt-VBR Class C) in Mbps	A	1.75	1.49	1.32	1.3
% of 22 sources at PCR (5918 cells 1.866 Mbps)	B	93.8%	79.8%	70.7%	69.7%
Effective bandwidth CLR 1.0 e-5 30 sources in Mbps	C	2.42	2.04	1.75	1.59
(A) as % of (C)	D	72.3%	73%	75.4%	81.8%
(A) as % of 30 sources at 2.544 Mbps (PCR)	E	68.8%	58.6%	51.9%	51.1%
Effective bandwidth CLR 1.0 e-6 (Nrt-VBR Class B) in Mbps	A	1.94	1.66	1.49	1.39
% of 24 sources at PCR (6456 cells 2.035 Mbps)	B	95.3%	81.6%	73.2%	68.3%
Effective bandwidth CLR 1.0 e-6 30 sources in Mbps	C	2.44	2.11	1.83	1.66
(A) as % of (C)	D	79.51%	78.7%	81.4%	83.7%
(A) as % of 30 sources at 2.544 Mbps (PCR)	E	76.3%	65.3%	58.6%	54.6%
Effective bandwidth CLR 1.0 e-7 (Nrt-VBR Class A) in Mbps	A	2.03	1.77	1.58	1.47
% of 25 sources at PCR (6725 cells 2.120 Mbps)	B	95.8%	83.5%	74.5%	69.3%
Effective bandwidth CLR 1.0 e-7 30 sources Mbps	C	2.46	2.16	1.9	1.72
(A) as % of (C)	D	82.5%	81.9%	83.2%	85.5%
(A) as % of 30 sources at 2.544 Mbps (PCR)	E	79.8%	69.6%	62.1%	57.8%

queue can extend to 1,000 cells would be 2.42 Mbps (Row C), which is almost equivalent to the *Constant Bit Rate* (CBR) of 2.54 Mbps.

The saving in applying the CAC to 22 rather than 30 sources is shown in Row D, where the effective bandwidth for 22 sources is represented as a percentage of the effective bandwidth for 30 sources. A guide to the interpretation of the results of Table 9-10 is given below in Table 9-11.

TABLE 9-11 Key to Results in Table 9-10

A	Effective bandwidth provisioned after 30 sources have been reduced by the mathematical probability method (reduction to 22 sources for CLR 1.0 e-5, 24 sources for CLR 1.0 e-6, and 25 sources for CLR 1.0 e-7).
B	The result in (A) as a percentage of the number of sources provisioned according to the mathematical probability method at PCR—that is, the % reduction in bandwidth attributable to the application of the CAC algorithm for calculation of the effective bandwidth.
C	Effective bandwidth provisioned for 30 sources with no prior adjustment for silence removal.
D	The result in (A) as a percentage of the effective bandwidth required for all the sources shown in (C).
E	The result in (A) as a percentage of the bandwidth of 30 sources at PCR—in other words, this would apply if the traffic were placed in a CBR QoS class.

9.10 Analysis of Results of Table 9-10

Row A of Table 9-10 shows the results of reducing the number of sources allowed in a talkspurt concurrently and then applies a CAC algorithm using the reduced traffic descriptors. (The original number of sources is reduced according to the mathematical probability method described in Section 9.8.3—the amount of the reduction in bandwidth directly attributable to the application of the CAC is shown as a percentage in Row B of Table 9-10.)

The results in Row B show that for the smallest buffer size of 1,000 cells, the effect of the CAC is to further reduce the bandwidth by between 4 (CLR 1.0 e-7) and 6 percent (CLR 1.0 e-5). For a buffer sized at 5,000 cells, the application of the CAC results in a further reduction in bandwidth of approximately 20 percent. For a buffer sized between 10,000 and 15,000 cells, the application of the CAC results in a reduction of approximately 30 percent.

Row C of Table 9-10 shows the bandwidth that would have been required if the CAC algorithm had been applied to all 30 sources, that is, without first reducing the traffic descriptors for silence removal.

Row D then compares the effective bandwidth for the number of sources reduced for silence removal (Row A), as a percentage of the result when the CAC process is applied to all 30 sources.

The results in Row D show that application of the mathematical probability method to take account of silence removal has reduced the effective bandwidth to between 72.3 (CLR 1.0 e-5 and buffer of 1,000 cells) and 85.5 percent (CLR 1.0 e-7 and buffer of 15,000 cells) of the effective bandwidth that would be provisioned by the CAC if the number of sources was not reduced. The scale of the reduction depends on the CLR target. The weaker the CLR, the better the reduction. Also, the buffer/queue size is a factor, and the scale of the reduction tails off as the buffer size is increased. For example, for a higher CLR of 1.0 e-7, there is very little difference between the gain achieved for a buffer of 1,000 cells (82.5 percent) and a buffer of 15,000 cells (85.5 percent). In other words, the bandwidth "bought" by allowing the buffer queue size to increase, decreases as the buffers are extended in size, and the first 1,000 cells of buffering buys most of the statistical gain. (For a CLR of 1.0 e-5, the range is 72 to 82 percent for 1,000 to 15,000 cells. For a CLR of 1.0 e-6, the range is 80 to 84 percent for 1,000 to 15,000 cells.)

Row E of Table 9-10 presents the results from Row A as a percentage of the original bandwidth required for 30 sources at PCR. The results show the bandwidth requirements reduced by silence removal and application of the CAC as a percentage of the bandwidth used if voice traffic were placed in a CBR class, that is, with no prior reduction for silence removal and no statistical gain. The total reduction available by reducing the number of sources for silence removal and applying the CAC algorithm to calculate effective bandwidth ranges from 79.8 (queue size 1,000 Nrt-VBR Class A 1.0 e-7) to 51.1 percent (queue size 15,000 Nrt-VBR Class C 1.0 e-5) of the CBR equivalent for an unreduced number of sources.

9.11 The Effect of Cell Loss on Voice Quality

When a number of channels are contributing to an aggregate flow and the channels are sending samples on a round-robin basis as in AAL2, we can examine the case where cell loss occurs because of a buffer overflow and look at how often this would happen based on a target CLR.

This raises the following question: Given a selected CLR target, how long is the interval, in seconds or minutes, before a cell might be lost?

The total cell rate of a number of channels combined in a single *Virtual Channel Connection* (VCC) depends on several factors. These are

- The number of channels contributing to the VCC
- AAL2 CU-timer setting

- The degree of compression selected by the voice coder selected
- Whether silence suppression is selected or not

If a number of different channels are contributing to the same virtual connection and cell loss occurs, in all probability, it would not strike the same channel twice in succession. However, we should look at the worst-case hit rate for cell loss and examine the possibility that cell loss does affect the same channel more than once, as well as the average-case hit rate for cell loss over the aggregate flow of an entire VCC.

As the number of channels contributing to the VCC is increased, the likelihood of cell loss repeatedly affecting the same channel will decrease. A CLR target at the queuing point where the voice channels are adapted to ATM that is weaker than that for the end-to-end CLR may be appropriate according to the type of voice services being combined in the VCC.

Subjective studies performed on 64-Kbps PCM and 32-Kbps ADPCM voice indicate that a weak packet loss target of 1.0 e-2 would not cause a perceptible loss of voice quality [55]. Therefore, another approach to optimizing bandwidth might be to permit a less challenging CLR at the queuing point, and this would permit a more aggressive reduction in the bandwidth to be provisioned.

Example There are 100 channels carrying voice, contributing to a single AAL2 VCC. Using the methods described in this book, it is estimated that bandwidth equivalent to what would be required by 66 sources should be provisioned for when sizing the PCR and MBS traffic descriptors in order to achieve a target CLR of 1.0 e-9.

The sources are using a G.726 32-Kbps coder with an average bandwidth per channel of 41.5 Kbps. This is a coder producing a constant bit rate when the source is in a talkspurt. Over time, the output rate of the coder will switch back and forth between the talking rate and the silence rate.

Applying a CLR target of 1.0 e-9, it can be calculated that the average gap between cell loss occurrences, affecting any of the aggregate cell flow of 100 channels, will be 28 hours. This figure can represent the shortest worst-case interval between errors, that is, if all of the cell losses were applied to a single channel. If the cell loss were evenly spread over all the 100 channels, a single channel would only see an error every 2,810 hours on average, which is every 117 days—in other words, a very long phone call!

If a weaker CLR target of 1.0 e-5 is applied, it is estimated that reduced bandwidth that is equivalent to 57 sources should be provisioned for when sizing the PCR and MBS. The shortest worst-case interval between cell losses over the aggregate flow of 100 channels increases to a cell loss every 10 sec. The average interval between cell losses evenly spread over all the 100 channels is 16.86 min.

If a weaker CLR of 1.0 e-2 is applied, it is estimated that bandwidth requirements can be further reduced, and only 47 sources need to be provisioned for when sizing the PCR and MBS traffic descriptors. The shortest worst-case interval between cell losses if they were all to occur in one channel now equates to a single cell loss every 10 ms. The average interval between cell losses if they were evenly spread over all the channels is 1 sec.

Note that although conversational voice may be reasonably tolerant of cell loss, other voice services (such as fax) that may intermittently use the same channel as the voice call may not be so tolerant. Also note that the previous examples are illustrated for the loss of a single cell, whereas in reality, cell losses are likely to occur in bundles, and therefore proportionately less frequently.

Conclusions

- Mathematical modeling is used to establish the probability that if a number of sources are simultaneously in a talkspurt, they do not all overlap for the whole duration of a talkspurt. The modeling assumes a method of *equivalent bursts*, which means that the total traffic stream is a superposition of an infinite number of sources generating bursts of fixed height and exponential length, which arrive according to a Poisson process.
- Setting of the PCR and MBS traffic descriptors is reduced to the number of sources estimated to remain in a talkspurt for the whole duration of an average estimated talkspurt.
- The estimates of PCR and MBS will depend on the number of sources estimated to be initially in a talkspurt concurrently. This may be all contributing sources or a reduced number of sources.
- The sizing of the estimated PCR and MBS traffic descriptors will depend on the number of cells that would be produced in such an average-length talkspurt, and this will depend on the type of coder in use and the packing factor (as in AAL5) or timer setting (as in AAL2).
- When a CAC algorithm is applied an increased reduction in the effective bandwidth is possible when,

- PCR and MBS are set according to the number of sources likely to continue in a talkspurt.
- SCR is set to average gain at 0.45 of PCR.

- The delay sometimes experienced by a connection constrains the length of the buffer queue.
- In addition to a CLR target the Rt-VBR classes must also meet the CDV constraint of 250 ms (Rt-VBR Class A) and 2,500 ms (Rt-VBR Class B). This severely constrains the viable queue lengths that can be tolerated by a connection in those classes, and hence very little statistical gain is realized through the application of the CAC algorithm.
- The CAC algorithm will generally not allow a further reduction in the provisioned PCR bandwidth within Rt-VBR Class A, with the exception of when a higher-bandwidth trunk is used (for example, STM-1) and a reduction in that case is only around 1 percent of the PCR bandwidth. This is seen to be due to the fairly high MBS, which results when a number of sources are in a talkspurt concurrently. In calculating effective bandwidth, the effect of the MBS dominates over the SCR.
- The mathematical method of calculating the traffic descriptors for silence-suppressed voice prior to applying the CAC demonstrates that there is a scaleable way to provision bandwidth for voice services, which is shown to be consistent with the maintenance of the CLR target. Without a prior reduction in the traffic descriptors, for example, through applying the mathematical methods described here, voice services in the Rt-VBR classes would not benefit on any significant scale from the additional statistical gain due to silence removal.
- The sizing of the buffers for voice in a VBR Class of Service is shown to be the strongest factor in achieving statistical gain. The greatest part of the gain realized after reducing the number of sources for silence removal using the mathematical probability method described here, prior to applying the CAC, was achieved when the buffers were allowed to extend to 1,000 cells (Row D, see Table 9-10). The increasing of the buffer queue length to 5,000, 10,000, and 15,000 cells saw a tailing off of the percentage gain. Buffering buys more statistical gain when a lower CLR is used, and the first 1,000 cells of buffering buy most of the gain available when even larger buffers are used.
- The methods developed here for adjusting the traffic descriptors are also useful where one requires a CLR that is lower or higher

than the CLR normally applied to the selected QoS class. For example, in applying the mathematical probability method to an Rt-VBR class, for estimating a reduction due to silence removal, it would be possible to apply the reduction applicable for a CLR target lower than 1.0 e-9.[4] This would be a case where the network user would like to have the voice traffic prioritized higher than the Nrt-VBR traffic on the ATM switch but prefers to risk undersizing the MBS (with a slightly greater risk of higher delay) in order to have greater throughput. Likewise, it may also be useful to apply the reduction applicable for CLRs higher than 1.0 e-7 to an Nrt-VBR class. For example, where all frame relay traffic is carried over an ATM backbone as Nrt-VBR, it would be natural to require a higher CLR specifically for the voice traffic component.

[4] Table 8-1 shows a recommended reduction for all target CLRs, irrespective of the QoS class, for a talkspurt sized at 1.34 sec.

Chapter

10

Queuing and Shaping

Analysis must determine the relationship between the potential statistical gain associated with a *Cell Loss Ratio* (CLR) target, the queuing delay, and delay variation that are experienced at a queuing point that will be present on the ingress linecard.

This chapter explains where delay due to queuing and shaping can occur in network hardware. Queue length distributions can be obtained in order to predict buffer requirements. The queue length can be traded off against potential statistical gain. A simulation tool is employed in order to observe the effects on the queue delay when the average talkspurt size is varied.

10.1 Sources of Delay in Voice Networks

Voice is a very distinct type of traffic. A voice connection is very sensitive to delay, and core networks should be designed with this traffic characteristic in mind. Factors contributing to the delays experienced by voice services can be caused by the following:

- **Ingress processing delay** This may include:
  - **Compression time** The time taken to compress a voice channel down from a 64-Kbps *Pulse Code Modulation* (PCM) channel to a low-bit-rate channel
  - **Assembly time** The time taken to fill the *Asynchronous Transfer Mode* (ATM) cell with PCM or compressed voice
- **Traffic shaping delay** This occurs when ingress cells are delayed at the shaper's queue before being transmitted into the ATM network. Cells are delayed in the shaper queue in order to

ensure that a connection's traffic descriptor is not violated. Cells in a *Constant Bit Rate* (CBR) *Quality of Service* (QoS) class are shaped to *Peak Cell Rate* (PCR). Cells in a *Variable Bit Rate* (VBR) QoS class are shaped to PCR and *Sustainable Cell Rate* (SCR).

- **Network delay** The time taken to transmit the cell from one end of the ATM network to the other. This may include:
  - Queuing time within the switch
  - Propagation delay across the interswitch links
- **Egress processing delay** This may include:
  - Time taken to decompress the voice channel back to 64 Kbps at the receiver
  - Delay caused by buffering to accommodate jitter that may occur in the network

The ATM Forum [16] has recommended a bounding method using a moment-generating function in conjunction with the Markov inequality method (see Appendix D—Markov's Inequality Algorithm [69]). This method is known as the *Chernoff method*. An important advantage of this method is that it allows the computation of the total CDV by using the mean and variance of the delay in switches.

10.2 Relationship Between CLR and Queue Delay

Methods of deriving the effective bandwidth for ON/OFF sources through application of a *Call Admission Control* (CAC) Algorithm usually consider the CLR target and the buffer (queue) size. Table 8-1 (average talkspurt = 1.34 sec) defines the number of sources likely to be in a talkspurt concurrently based on a CLR target.

Reduced traffic descriptors—PCR, SCR, and *Maximum Burst Size* (MBS)—are input into the CAC Algorithm and mapped to an effective bandwidth, αi, which satisfies the QoS constraints. The effective-bandwidth approach views each virtual connection in isolation, that is, as if it were alone at the queuing point.

The CAC rule for admitting new connections compares the sum of the existing connections with the overall link capacity:

$$\sum_{i=1}^{n} \alpha_i \leq \textit{Available Capacity} \tag{10-1}$$

where n = all the connections including the new one

Statistical gain occurs when the effective bandwidth provisioned is less than the PCR.

10.3 Delay Due to Queuing and Shaping

The egress buffer on the ingress linecard holds the queue for each virtual connection carrying voice from one or several channels. The cell removal rate from the queue, or *service rate*, will be controlled by a traffic shaper that enforces the cell flow emitted as shown in Figure 10-1. The maximum tolerable queue length depends on the maximum tolerable additional delay that would be incurred due to the traffic shaping at this point.

For example, in the simplest case where the output rate of the shaper has a CBR value of R cells/sec, if a tolerable delay is T seconds, the maximum queue depth should be TR cells. When a VBR service is selected, shaping takes place based on both the PCR and SCR traffic descriptors.

Concurrent talkspurts arrive in a burst. Within a burst of cells, described by the MBS parameter, there can be several talkspurts.

During a talkspurt, a single source will produce a periodic stream of cells. When AAL5 is being used, between 31 and 537 cells can be generated in a single talkspurt depending on the coder type (refer to Table 9-1). When AAL2 is used, talkspurts from more than one source share the same cell stream. For example, two sources will create a *double-size* talkspurt (refer to Table 9-2).

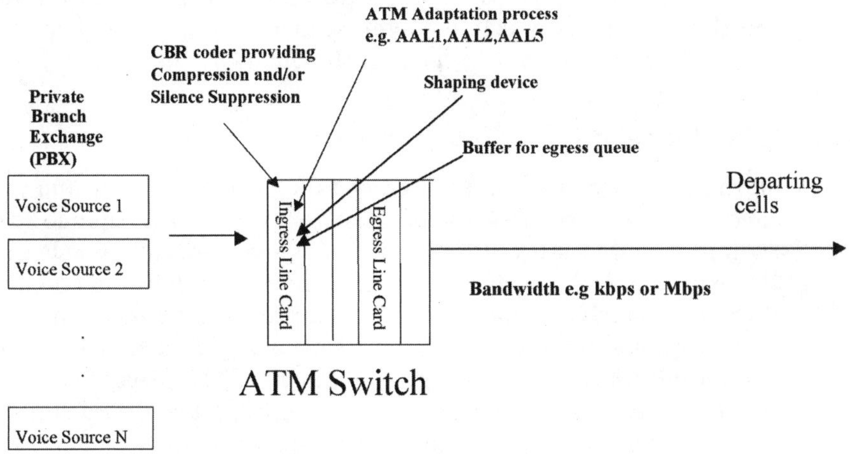

Figure 10-1 Controlling the removal rate from the egress buffer

A burst is either outside the system and arriving, waiting in the system in the queue, or being served, that is, delivered from the queue. During a burst, the queue will be emptied at the PCR. When the burst has expired, the queue will be emptied at the SCR. The SCR is considered to be the long-term cell-emission rate and therefore should reflect the average saving due to silence suppression. The traffic rate on the connection will drop below the SCR during periods of silence.

If a burst is longer than expected and exceeds the MBS descriptor, the remainder of the burst will be played out of the buffer at the lower SCR rather than at the PCR. Thus, the sizing of the MBS incurs both the risk of cell loss and the risk of additional delay. These risks should be weighed seriously against the potential statistical gain offered when a VBR service is selected rather than a CBR service.

10.4 Bursting the Buffer Queue

The probability of cell loss has already been found by totaling the probabilities of all the burst states that are greater than the mean burst size (based on an average talkspurt of 1.34 sec). A sum of probabilities was established in order to determine the probability that an MBS would be exceeded where the burst size was based on the number of sources expected to be concurrently in a talkspurt and a priori flow characteristics.

If too many sources are in talkspurt simultaneously and burst the queue, the ensuing cell loss will occur at the queuing point in the egress buffer of the ingress linecard. It is possible to trade off the probability of cell loss against increased queuing delay by controlling the maximum depth of the queue on the card.

10.5 Buffer Modeling

If the combined effect of the average talkspurt arrival rate and the average talkspurt size exceeds the service capacity, the queue size will grow indefinitely. If the buffer is undersized, cell loss should be expected. If the buffer is oversized, increased delay will occur.

Several stochastic models have been proposed for modeling the egress queue. To assess the queuing behavior of packetized voice communication systems, Daigle and Langford [27] discussed three models: a *semi-Markov process* (SMP) model, a *continuous time Markov chain* (CTMC) model, and a *uniform arrival and service* (UAS) model. Numerical results from each of the three models are compared to each other, to results obtained from a discrete event simulation program, and to results obtained from an M/D/1[1] analysis. Results reveal that the SMP, CTMC, and UAS models all predict longer queue lengths than those showed in the simulation.

However, it has also been shown that queue length distributions obtained from the SMP and UAS models are reasonable for engineering purposes. Such queue length distributions might be used, for example, to predict the buffer capacity needed to ensure that the fraction of packets lost due to buffer overflow is less than some maximum value.

Based on the results of their numerical work, Daigle and Langford stated that it did not appear that the M/D/1 model would provide useful results unless the number of voice sources was large, say, in the hundreds.

The M/M/1 queue with exponential size talkspurts and exponential arrival rates is considered here to fit the VBR voice model better than the M/D/1 model, which is considered to be more appropriate to traffic in the CBR class.

The VBR voice traffic is being shaped to PCR and SCR, which reflect the worst- and average-case bandwidth, respectively.

Kleinrock [47] shows that when the mean normalized queuing time for the M/M/1 and M/D/1 systems are compared, the system with constant service time (M/D/1) would have half the average waiting time of the M/M/1 system with an exponentially distributed service time.

Also, Daigle and Langford's [27] mathematical models produced queue lengths longer than those obtained from a simulation. This may be because of the restricted buffer size. It is, therefore, considered preferable to use an infinite buffer size in the simulation investigations in order to observe the queue behavior freely.

10.5.1 Steady State

When a queuing system is first put into operation, and for some time afterwards, the length of the queue depends strongly upon the number of customers queued up waiting and the length of time the system has been in operation. The system is said to be in the *transient* state. After the system has been in operation for a long time, the influences of the initial conditions and of the time since startup have "damped out," and the number of customers in the system and queue is independent of time; the system is in the

---

[1] M/D/1 stands for memoryless interarrival time/deterministic service time/single server. Further explanation of queuing theory notation is provided in Section 7.3.

steady state. Therefore, both the average- and worst-case statistics on queue delay are affected by ramping up the queue.

10.6 Opnet Simulation

To discover the effect of the talkspurt size on the delay in the queue, Opnet[2] simulation of an M/M/1 queue was used to capture the dynamics of the buffer for a variable aggregate flow through simulation. It is possible to observe both the shorter time scale burstiness of the traffic prior to reaching a steady state, as well as to capture the variation in the flow rate of the aggregated traffic over a longer time scale.

10.6.1 The M/M/1 Queuing Model

The M/M/1 queue consists of a *first in, first out* (FIFO) buffer with talkspurt cells arriving randomly according to a Poisson process and a processor called a *server* that retrieves cells from the buffer at a specified service rate.

Mean talkspurt arrival rate, mean talkspurt size (service requirement), and service capacity are represented by λ, $1/\mu$, and C, respectively. The behavior of an M/M/1 queue was simulated using Opnet Version 6.0.

10.7 The Opnet Queuing Model

The Opnet simulation follows the model shown in Figure 10-1. At the ingress port of an ATM switch, a voice services linecard receives traffic from the voice source and generates ATM cells that contain

- **Active data**
- **Filler in the AAL frame** Has been incurred because it has not been possible to completely fill the frame to meet with the delay requirements of the rest of the network
- **Framing overheads** Including AAL5 overheads and ATM cell header

[2] Opnet is a product of MIL 3 Inc., 3400 International Drive NW, Washington D.C., 20008.

- **Silence indication overheads** Short packets to preserve the timing of the connection
- **Interferor cells**

The main source of interferor cells over a connection normally operating between linecard endpoints are the CAS heartbeat cells. Normally, a heartbeat packet of 1 cell would be sent every 5 sec. This is a keepalive mechanism to detect if the ATM link is still active. This cell would be sent during both talkspurts and silence.

Another source of interferor cells is OAM Alarm Surveillance, or OAM Connectivity Verification cells, although these cells are not present on a normally operating connection. Within the bandwidth calculations, one additional cell/sec of bandwidth is allowed (above and beyond the other bandwidth requirements of the connection) for interferor cells.[3]

Opnet modeling is represented by three separate configurations of a traffic generator, as shown in Figure 10-2:

- Voice cells from n sources generating exponentially sized talkspurts with exponential interarrival times that have been encapsulated in an FRF.11 and AAL5 frame. The number of sources is varied from 9 to 30 in separate simulations.
- Interferor cells from n sources.
- Signaling cells (CAS) from n sources (sized at 50 cells/sec constant rate).

10.8 Opnet Results

Opnet simulations are carried out for talkspurt sizes of 269 and 135 cells, equivalent to G.726 PF = 1 (AAL5 framing) and G.726 PF = 4 (AAL5 framing), respectively. The objective of the simulation is to observe the effect of a different talkspurt size (in cells) on the queue behavior as the number of sources sending talkspurts is varied from 9[4] to 30. Figure 10-3 illustrates that the

[3] Interferor cell characteristics are independent of other factors affecting connection bandwidth (for example, coder and packing factor)—that is, the amount of bandwidth allocated for interferor cells is a constant and is not affected by the selected coder and packing factor.

[4] Reliable statistical gain due to silence removal cannot be established by the method described in the research for the range of CLRs of interest for less than 9 or 10 channels.

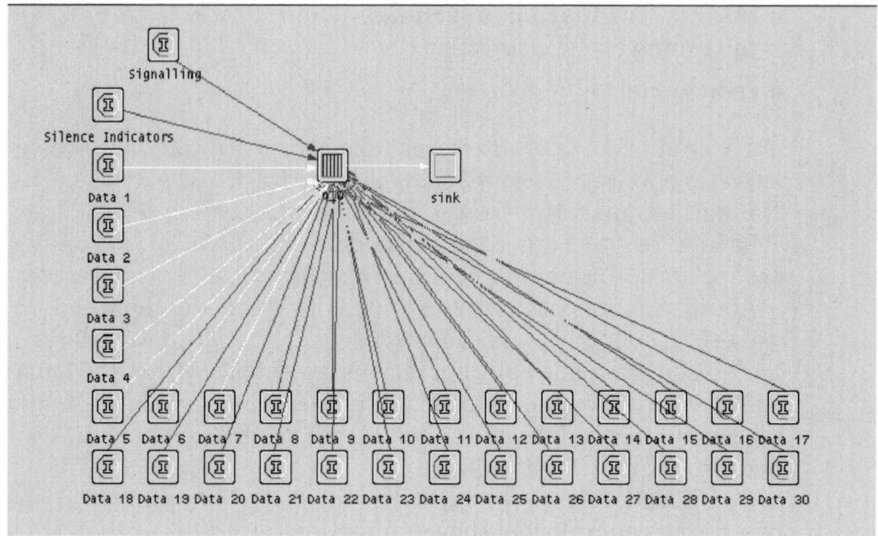

Figure 10-2 Traffic generators for voice, signaling and silence indicators used in Opnet

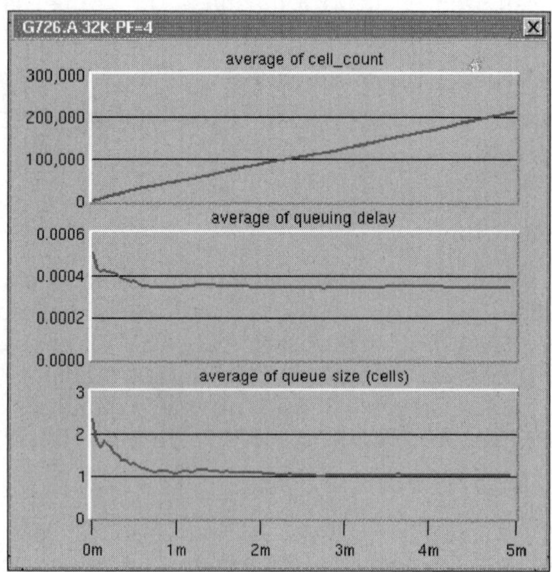

Figure 10-3 Averages of cell count, queuing delay, and queue size for G.726 32-Kbps PF 4 (m = minutes)

average simulation takes just about one minute to reach a steady state. The results for a talkspurt, sized at 135 cells, show that the queue oscillates around zero (as expected). The delay is fairly constant (after steady state is reached). The cell count is recorded in order to estimate the actual bandwidth used per second.

10.9 Calculation of Expected Average Delay per Talkspurt

The expected average delay experienced by a burst made up of talkspurts from several sources in the M/M/1 queue can be calculated as in the following example.

Example 1: G.726 32-Kbps Compression Coder, Packing Factor = 1, Using AAL5 In the ON state of the simulation, 30 sources send talkspurts[5] (TS) averaging 269 cells in length, which arrive on average every 3.01 sec (1.34 sec of talk and 1.67 sec of silence).

$$\text{The Mean Service Requirement for one TS} = \frac{1}{\mu} = 269 \text{ cells}$$

$$\text{Mean arrival rate for 30 sources: } \lambda = \frac{30 \text{ TS}}{\textit{Mean Interarrival time}} = \frac{30}{3.01} = 9.9668 \text{ TS/sec}$$

Service capacity: C = 353,207 cells/sec where the speed of the egress line is assumed to be STM-1, and a talkspurt burst is assumed to be emptied at PCR.

$$\text{Mean Service Rate: } \mu\text{C} = \frac{1}{269}\,353{,}207 \cong 1{,}313.037 \text{ TS/sec}$$

$$\text{Mean delay: W} = \frac{1}{\mu C - \lambda} = \frac{1}{1{,}313.037 - 9.9668}$$

$$= \frac{1}{1{,}303.0702} = 0.0007674 \text{ sec}$$

Results for the Opnet simulation illustrated in Figure 10-4 show an average delay of 0.7076 ms, but a simulation of sufficient duration would produce the expected results.

[5] CAS signaling and silence indicators are not included in Example 1 or Example 2.

Example 2: G.726 32-Kbps Compression Coder, Packing Factor = 4, Using AAL5 In the ON state of the simulation, 30 sources send talkspurts (TS) averaging 135 cells in length, which arrive on average every 3.01 sec.

$$\text{The Mean Service Requirement for one TS} = \frac{1}{\mu} = 135 \text{ cells}$$

$$\text{Mean arrival rate for 30 sources: } \lambda = \frac{30 \text{ TS}}{\textit{Mean Interarrival time}} = \frac{30}{3.01} = 9.9668 \text{ TS/sec}$$

Service capacity: C = 353,207 cells/sec (STM-1)

$$\text{Mean Service Rate: } \mu C = \frac{1}{135} 353{,}207 \cong 2{,}616.3481 \text{ TS/sec}$$

$$\text{Mean delay: W} = \frac{1}{\mu C - \lambda} = \frac{1}{2{,}616.3481 - 9.9668}$$

$$= \frac{1}{2{,}606.3813} = 0.0003837 \text{ sec}$$

Table 10-1 shows the results of the Opnet simulation for up to 30 sources using a G.726 32-Kbps compression coder and AAL5 framing with a packing factor of 1, producing a talkspurt sized at 269 cells, and a packing factor of 4, producing a talkspurt sized at 135 cells.

Figure 10-4 illustrates the Opnet simulation results given in Table 10-1 for the average- and worst-case delay incurred in the queue for two talkspurt sizes—1.34 sec and 2.0 sec. The maximum size of a burst (in talkspurts) determines the delay experienced in the queue where the size of the burst is a product of the number of sources and the talkspurt size:

- The worst-case delay is approximately twice the average delay. This is intuitively correct—that is, approximately half of the time, channels will be in a talk state and the remainder of the time in silence. A worst case occurs when all of the channels are in talk together. For example, when 10 sources are sampled and each source is potentially sending a talkspurt sized at 269 cells, the total worst-case burst is 2,690 cells if all the sources are talking at the same time. As shown in Table 10-1 and illustrated

TABLE 10-1 Opnet Simulation Results on Queuing Delay for G.726 PF = 1 and PF = 4

Opnet Simulation Results on Delay for a Talkspurt (Appropriate to a G.726 Coder Using AAL5 with a Packing Factor = 1 and Packing Factor of 4) an STM-1 Egress Trunk = 353,207 Cells/sec, Simulation Time = 300 seconds, Number of Silence Indicator Cells Sent During an Average Silence Period = 33

	Packing Factor = 1 (269 cells)		Packing Factor = 4 (135 cells)	
Number of Voice Sources	Average Delay in Queue (ms)	Worst-Case Delay in Queue (ms)	Average Delay in Queue (ms)	Worst-Case Delay in Queue (ms)
9	0.5604	1.3708	0.2997	0.7066
10	0.5855	1.2784	0.3067	0.6588
11	0.5813	1.2075	0.3096	0.6221
12	0.5879	1.1636	0.3117	0.5989
13	0.6065	1.1636	0.3201	0.5989
14	0.6054	1.1636	0.3186	0.5989
15	0.6208	1.2492	0.3255	0.6400
16	0.6220	1.2492	0.3270	0.6400
17	0.6460	1.2492	0.3316	0.6400
18	0.6383	1.2863	0.3363	0.6578
19	0.6521	1.2433	0.3361	0.6355
20	0.6405	1.2033	0.3441	0.6148
21	0.6558	1.2033	0.3333	0.6148
22	0.6574	1.1742	0.3505	0.5987
23	0.6547	1.1939	0.3518	0.6082
24	0.6684	1.2298	0.3426	0.6258
25	0.6753	1.2298	0.3489	0.6258
26	0.6889	1.1829	0.3431	0.6017
27	0.6735	1.1773	0.3512	0.5983
28	0.6981	1.1471	0.3462	0.5828
29	0.6952	1.1083	0.3471	0.5629
30	0.7076	1.1469	0.3566	0.5820

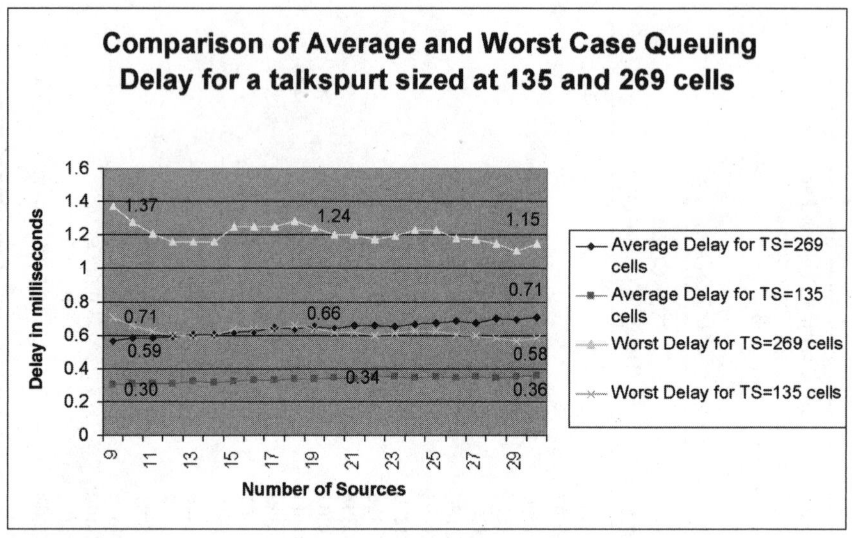

Figure 10-4 Delay in queue for two different talkspurt (TS) sizes and 9 to 30 sources

in Figure 10-4, this connection experienced an average delay of approximately 0.6 ms and a worst-case delay of 1.3 ms.

- When the same number of voice sources are contributing, the average- and worst-case delays halve when the talkspurt size is halved. Note the average delay for 20 sources sending talkspurts half the size at 135 cells (total also 2,690 cells if all are talking) is 0.3 ms, and the worst-case delay is 0.6 ms.
- Delays remain fairly constant. As the number of sources is increased, the average delay increases slightly and the worst-case delay decreases slightly.

When a number of voice samples are combined into the same VCC, or VP, there are three factors in determining that the delay incurred:

- **The talkspurt size** Determines the size of a burst from several sources concurrently in a talkspurt. The higher the compression rate, the smaller the burst of cells arriving in the buffer.
- **The number of sources contributing** The greater the number of sources contributing, the smaller the gap between the arrival of talkspurts.

- **The service rate of the egress buffer** Determines how quickly talkspurts can be removed from the buffer.

10.10 Opnet Results on Bandwidth

Table 10-2 shows the comparison of the results of a cell count taken during simulation of 30 voice sources, where the estimated

TABLE 10-2 Effective Bandwidth Results Using the Mathematical Probability Tables for Silence Removal and Application of a CAC Algorithm

Result	Description	Number of Sources by Mathematical Probability Method	Total Bandwidth After Adjustment Silence Removal TS = 269 cells (Kbps)	After the CAC Algorithm Is Applied to All 30 Sources	After Applying CAC to the Bandwidth Reduced for Silence Removal
Column 1	Column 2	Column 3	Column 4	Column 5	Column 6
Row 1	Simulation results TS = 1.34 sec	NA	1176	NA	1176
Row 2	45% of all sources at PCR	NA	1039	NA	1039
Row 3	CLR at 1.0 e-4 (10,000 cells)	**21**	**1891**	1762	***1375***
Row 4	CLR at 1.0 e-5 (10,000 cells)	**22**	**1981**	1860	***1445***
Row 5	CLR at 1.0 e-6 (10,000 cells)	**24**	**2158**	1942	***1580***
Row 6	CLR at 1.0 e-7 (10,000 cells)	**25**	**2247**	2012	***1680***
Row 7	CLR at 1.0 e-8 (880 cells)	**26**	**2336**	2621	***2266***
Row 8	CLR at 1.0 e-9 (880 cells)	**27**	**2426**	2629	***2363***
Row 9	CLR at 1.0 e-10 (880 cells)	**28**	**2514**	2636	***2457***
Row 10	CLR at 1.0 e-11 (880 cells)	**28**	**2514**	2640	***2462***
Row 11	No savings allowed for silence	NA	2692	2692	2692

Columns 4 to 6 of Rows 2 to 11 (inclusive) include an allowance for the overhead due to silence indicator cells of 5% of the aggregate bandwidth, plus an allowance of 21.2 Kbps per connection for CAS signaling.

bandwidth requirements, which have been reduced to allow for silence removal, according to the mathematical probability method described in Chapter 8, and the overbooking recommendations given in Table 8-1.

Results are shown for a G.726 32-Kbps coder using a packing factor of 1 and a packing factor of 4, which equates to a talkspurt sized at 269 cells and 135 cells, respectively.

10.10.1 Key to Table 10-2

Row (1) = Opnet simulation results Using Opnet version 6.0, 30 voice sources were simulated sending talkspurts of exponentially distributed length, sized according to G.726 PF = 1 (269 cells). Cells arrive as a burst with an exponentially distributed average interarrival time of 3.01 sec (1.34 sec of talk collected + 1.67 sec of silence). The average of the exponential service time is set to 353,207 cells/sec (STM-1) in all cases. Silence cells are sent at a rate of approximately 5% of the aggregate bandwidth. CAS is set at 50 cells/sec (21.2 Kbps) per source (constant).

Row (2) = expected average gain Bandwidth requirements based on average silence occurring (1.67/3.01 sec). An average gain of 55.48 percent due to silence suppression is expected, minus the allowance for silence indicators sent during silence, plus an extra allowance for CAS signaling.

Row (3) Column 3 (bold) Provisions a reduced number of sources as described in the probability tables for silence removal (Table 8-1). According to this method for 30 sources, bandwidth for 21 sources would be provisioned to achieve a risk of cell loss at 1.0 e-4. **Column 4 (bold)** shows the bandwidth required in Kbps for 21 sources. One interferor cell/sec is added to each source, and 50 cells/sec for CAS signaling are added to the aggregated sources. For example, a single channel using a G.726 32 Kbps coder and AAL5 framing with a packing factor of 1 requires 84.8 Kbps without silence removal. See Table 5-5. The addition of the silence indicator cells increases this to approximately 89.04 Kbps. Twenty-one channels would require a total of 89.04 × 21 = 1869.84 Kbps. CAS increases this by 21.2 Kbps to a total of 1891.04 Kbps.[6] **Column 5** shows the bandwidth required after a CAC Algorithm is applied using the unreduced traffic descriptors (that is, PCR/MBS config-

[6] Corresponding figures for G.726 with a packing factor of 4 are 41.72 Kbps, including the interferor cell.

ured for bandwidth for all 30 sources). **Column 6 *(bold and italic)*** shows the bandwidth required after a CAC Algorithm is applied using the traffic descriptors reduced according to the probability tables for silence removal, as described in Chapter 8 and Table 8-1. The queue is set to 10,000 cells, which could be appropriate for a CLR of this range without a CDV constraint.

Row (4) As Row (3) for a CLR of 1.0 e-5, bandwidth would be reduced to 22 sources. The queue length is set at 10,000 cells.

Row (5) As Row (3) for a CLR of 1.0 e-6, bandwidth would be reduced to 24 sources according to the probability tables for silence removal. The queue length is set at 10,000 cells.

Row (6) As Row (3) for a CLR of 1.0 e-7, bandwidth would be reduced to 25 sources according to the probability tables for silence removal. The queue length is set at 10,000 cells.

Row (7) As Row (3) for a CLR of 1.0 e-8, bandwidth would be reduced to the bandwidth equivalent to 26 sources according to the probability tables for silence removal. The queue length is set at 880 cells, which is appropriate for this CLR and an STM-1 egress trunk.

Row (8) As Row (3) for a CLR of 1.0 e-9, bandwidth would be reduced to the bandwidth equivalent to 27 sources according to the probability tables for silence removal. The queue length is set at 880 cells, which is appropriate for this CLR and an STM-1 egress trunk.

Row (9) As Row (3) for a CLR of 1.0 e-10, bandwidth would be reduced to the bandwidth equivalent to 28 sources according to the probability tables for silence removal. The queue length is set at 880 cells, which is appropriate for this CLR and an STM-1 egress trunk.

Row (10) As Row (3) for a CLR of 1.0 e-11, bandwidth would be reduced to the bandwidth equivalent to 28 sources according to the probability tables for silence removal. The queue length is set at 880 cells, which is appropriate for this CLR and an STM-1 egress trunk.

Row (11) Shows the case where there is no allowance for silence suppression at all—that is, this is the bandwidth equivalent to 30 sources at PCR (equivalent to CBR).

Figure 10-5 illustrates the results from Table 10-2. The endpoints of each curve are from the simulation results.

The results in Figure 10-5 illustrate how the bandwidth requirement climbs steeply at a CLR target of 1.0 e-8. This is

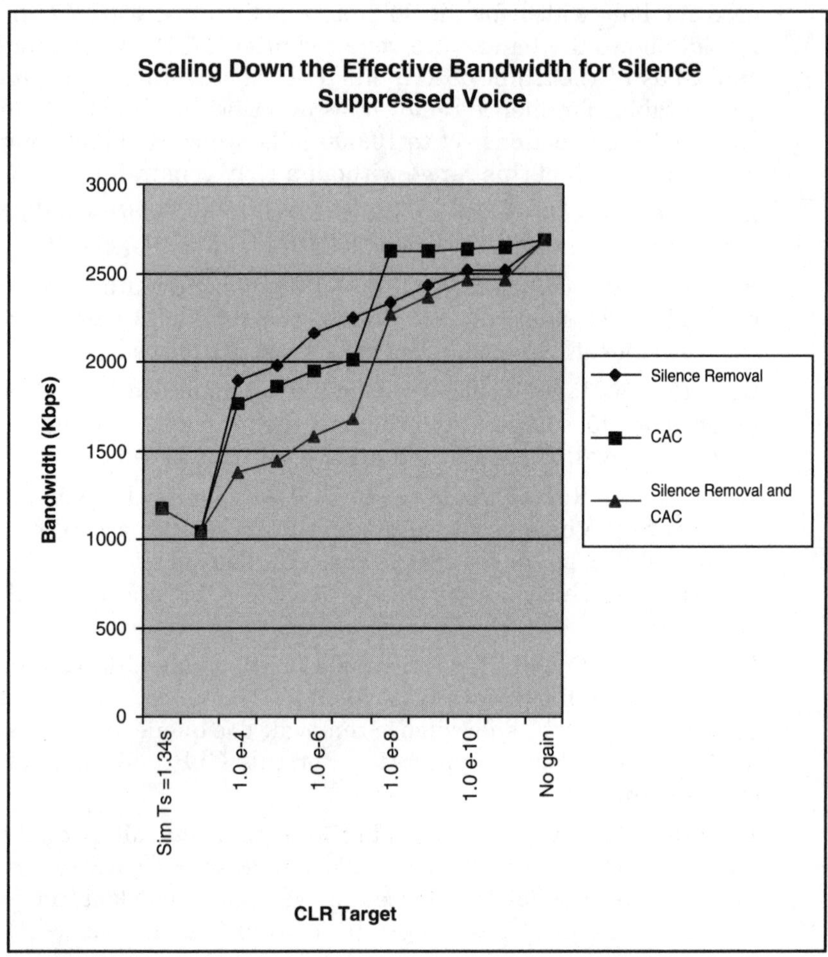

Figure 10-5 Comparison of effective bandwidth using silence removal and application of a CAC Algorithm

where a smaller buffer is enforced to meet the CDV constraint of a Rt-VBR service. The buffer size has been reduced from 10,000 cells, which is appropriate for Nrt-VBR services without a CDV constraint, to 880 cells,[7] which is appropriate for a Rt-VBR Class B service.

---

[7] Refer to Table 9-5—these buffer sizes are appropriate to an STM-1 trunk type at the egress port.

Prior application of a reduction in the traffic descriptors to account for silence removal (Table 8-1) has smoothed that jump. The curve of the CAC result is maintained but scaled down gracefully, allowing more statistical gain for the Rt-VBR Class B service while maintaining the CLR guarantee.

10.11 Dividing Channels into More Than One Connection

Selection of a QoS class for a connection influences the buffer queue length. It has been shown in Chapter 9 how the size of the MBS, based on the number of channels in a talkspurt at the same time, directly affects the amount of effective bandwidth provisioned.

When AAL5 is applied, each source has its own VCC. When AAL2 is applied, several sources can share a VCC. It is worth discovering if the dividing of a number of channels over more than one connection, which would reduce the MBS, would increase the statistical gain.

Table 10-3 shows the effective bandwidth results for 30 channels using AAL5 framing for a G.726 32-Kbps coder and a G.729 8-Kbps coder. The probability tables recommend the provisioning of 26 channels for a CLR of 1.0 e-8 and 24 channels for a CLR of 1.0 e-6. Appropriate buffer sizes are used. The CAC Algorithm is applied firstly for all the channels and secondly for half the channels.

The results show how dividing the connection into two bundles has relatively little benefit. The bandwidth provisioned for half the channels is almost exactly 50 percent of that for all the channels. As the buffer is allowed to be longer, a small amount of gain becomes possible. This is most noticeable for a buffer length of 883 cells and above.

These examples serve to emphasize that the underlying behavior of the source is CBR in nature, and that the most significant factor in achieving statistical gain is the buffer length.

Conclusions

- The Opnet simulation provides an opportunity to observe the average behavior of the multiplexed voice sources in the presence of both estimates of talk and silence periods, and for talkspurts that vary in size (in units of cells) according to the compression coder in use.

TABLE 10-3 The Effect of Dividing 30 AAL2 Channels into Two Bundles for Two Different Coder Types

Coder type	Number of channels which bandwidth is provisioned for in aVCC (Total number of channels contributing)	Results of Dividing Channels Between More Than One VCC for an AAL2 Coder					
		Rt-VBR CLR 1.0 e-8 CDV 250 ms	Rt-VBR CLR 1.0 e-8 CDV 2500 ms	Nrt-VBR CLR 1.0 e-6 No CDV constraint	Rt-VBR CLR 1.0 e-8 CDV 250 ms	Rt-VBR CLR 1.0 e-8 CDV 2500 ms	Nrt-VBR CLR 1.0 e-6 No CDV constraint
		E1 egress trunk type			STM-1 egress trunk type		
		Buffer 1 cell	Buffer 11 cells	Buffer 120 cells	Buffer 88 cells	Buffer 883 cells	Buffer 10,000 cells
		Effective Bandwidth provisioned in Kbps for all channels					
G.726 32k Average bandwidth per channel subject to optimized timer = 11.73 Kbps	13 (15) Rt-VBR CLR 1.0 e-8 MBS[2] 1,889 cells	539	539		533	484	
	26 (30) Rt-VBR CLR 1.0 e-8 MBS 3, 776 cells	1079	1078		1073	1020	
	12 (15) Nrt-VBR CLR 1.0 e-6 MBS 1,743 cells			487			307
	24 (30) Nrt-VBR CLR 1.0 e-6 MBS 3,485 cells			985			662
G.729 8k Average bandwidth per channel subject to optimized timer = 41.50 Kbps[1]	13 (15) Rt-VBR CLR 1.0 e-8 MBS 582 cells	152	152		147	115	
	26 (30) Rt-VBR CLR 1.0 e-8 MBS 1,163 cells	305	304		299	258	
	12 (15) Nrt-VBR CLR 1.0 e-6 MBS 537 cells			131			82
	24 (30) Nrt-VBR CLR 1.0 e-6 MBS 1,073 cells			272			168

[1] See Table 5-3 for calculation of the average bandwidth requirements for AAL2

[2] See Table 9-2 for calculation of the MBS size

- The delay statistics from the M/M/1 queue simulation show that the worst-case delay is always approximately twice the average-case delay. This is as expected—that is, worst case is all sources talking; the average case is approximately half the sources talking.
- Opnet simulations confirm that the delay incurred in the queue is directly related to the size of a burst derived from the product of the number of sources contributing and the coder unit size. However, increasing the number of sources, potentially contributing to a burst, has much less effect on bandwidth, than an increase in a coder unit size which causes a potentially larger number of cells in a talkspurt.
- When AAL5 is used, talkspurts come from a single voice source, and their size depends on the coder type and packing factor. Each cell stream is a single Virtual Channel Connection (VCC) with its own traffic descriptors. Talkspurt sizes (in cells) are restricted by the range of the packing factor.
- When AAL2 is used, talkspurt sizes can be based on several voice sources combined in one VCC. Talkspurt sizes can grow very large because up to 248 sources can share the same interface.
- The probability tables for silence removal, described earlier in Table 8-1, look at the total number of sources contributing and establish the number of sources likely to be in a talkspurt simultaneously. The tables allow for sizing of the traffic descriptors when channels are combined in the same connection, prior to applying the Call Admission Control (CAC) Algorithm.
- When AAL2 is used, Table 8-1 can be used to determine the number of channels that should be waited for out of the total number of channels available. The setting of the AAL2 timer can be based on the number of channels likely to be in a talkspurt at the same time. When finding an optimum setting for the timer in AAL2, the number of channels recommended by the probability tables will be an upper bound. It may not be possible to put the number of channels recommended by the probability tables into the same VCC if low compression is being used resulting in large talkspurt sizes.
- The buffering demands of large talkspurts may cause queuing delay to increase beyond the tolerance of the Quality of Service (QoS) Class, particularly if Rt-VBR classes are in use, as is normally the case for voice traffic. For higher compression rates and small coder units sizes, there will be much more flexibility and higher potential statistical gain.

- A number of AAL2 channels may be divided into more than one connection. Division reduces the talkspurt size in direct proportion to the number of connections it is spread over (refer to Table 9-2), which brings the setting of the MBS down in direct proportion to the number of channels. This in itself effects only a small percentage difference in the achievable gain, which as described in Section 9.7, is mainly determined by the buffer queue. The queue in turn is constrained by the QoS guarantees of the service class.
- The gains from dividing connections into bundles are small as the underlying behavior of the coder is CBR.
- The gain attributable to dividing connections when the queue lengths are allowed to be longer is more noticeable when high-rate compression is used and coder units are smaller.

Chapter

11

Main Conclusions

Throughout this book, specific conclusions that draw attention to key results have been listed after each chapter. The objective of this final chapter is to present some main conclusions. The previous points described at the end of each chapter of the book are not repeated here.

The market is presently dominated by AAL5. Most carriers believe that a change of adaptation layer to AAL2 for voice is warranted. However, it has been proved that this is not necessarily the case. It may be possible to choose a different adaptation layer type and achieve results that are still almost optimal—that is, the advent of AAL2 has not necessarily excluded the use of the other adaptation layer types, such as AAL1 and AAL5.

When there are several channels, the use of AAL5 will always incur a higher delay than AAL2, but AAL5 will allow for easier interworking with other traffic types, such as frame relay.

AAL5 will potentially incur higher delay and *Cell Delay Variation* (CDV) at cell fill than AAL2. If a high packing factor is selected and silence occurs when only one or two coder units have been loaded into the AAL5 frame, the frame will still wait the duration of the remaining coder units. This delay will increase with a higher packing factor.

A similar effect can potentially be produced in AAL2 through the timer setting; however, the maximum additional delay is in general the duration of one coder unit, and when there are a number of channels contributing, this delay reduces proportionately. Hence, the delay and CDV at cell fill is usually lower when using

AAL2 than when using AAL5. This delay and CDV will decrease as the number of AAL2 channels is increased.

AAL5 allocates a separate virtual circuit to each voice channel, whereas AAL2 enables voice from several channels to be combined in one virtual circuit. AAL2 bandwidth gains exceed those offered by AAL5, but they are only realized when a number of circuits are traveling to the same destination exit point from the ATM network unless AAL2 subcell switching is available.

The application of a higher compression factor reduces the difference between the bandwidth requirements of AAL2 and AAL5. The further reduction of bandwidth due to silence suppression narrows the gap between the bandwidth gain offered by AAL2 and AAL5 even further.

Research described here into the mathematical modeling of voice channels has demonstrated a way to separate the effects of estimating the mean talk state and mean silence state when calculating the probability that a number of talkspurts will arrive together and overlap.

This means that it is possible to adjust the expected length of a talkspurt to the average 1.34 secs, or more pessimistically, to 2.0 secs, and adjust the traffic descriptors and overbooking rate accordingly.

Adjustment will be useful in various countries. In Asia, for example, the profile of voice characterization as described in terms of the duration of average talk and average silence periods is different from the profile in Europe or the United States.

Up to now, the statistical gain, due to silence suppression, has been estimated to be up to 50 percent, but this research has shown that it can be more precisely quantified. Furthermore, a reduction in the provisioned bandwidth can be tied to a risk of cell loss, which can be selected to maintain the performance parameters of the selected *Quality of Service* (QoS) class.

Overbooking is not normally recommended for the *Constant Bit Rate* (CBR) and Rt-VBR classes as it implies that the QoS parameters cannot be guaranteed. This research shows that there is strong evidence that these QoS targets can be maintained in the presence of silence suppression when a scaling factor is properly applied.

When Rt-VBR classes are selected, minimal statistical gain results under normal conditions due to silence removal. Buffer sizes must be restricted in order to maintain the CDV constraints of these QoS classes (particularly Rt-VBR Class A). The mathematical probability methods described here allow for a reduction of circa 13 percent for a *Cell Loss Ratio* (CLR) of 1.0 e-9 due to silence removal. The application of the *Call Admission Control* (CAC) may increase that figure by an additional 1 percent (E3) to 3 percent (STM-1) for Rt-VBR Class B. However, the CAC does not allow any further reduction in bandwidth for queue lengths appropriate to T1/E1 trunks or traffic in the Rt-VBR Class A.

Nrt-VBR classes offer the most statistical gain. It is important to note that the first 1,000 cells of buffering provide most of the available statistical gain, particularly for more stringent CLR targets. Reducing the traffic descriptors for silence removal *prior* to applying the CAC yields an effective bandwidth of approximately 70 to 80 percent of what would have been required otherwise.

Research has shown that voice traffic can tolerate a less stringent CLR than other types of traffic. Analysis of the average time interval between the cell loss implied by a selected CLR target indicates that a weak target may be appropriate for voice traffic. This result leads to a corresponding increase in the achievable statistical gain in bandwidth for silence-suppressed voice traffic.

Appendix A

FRF.11, FRF.8, and FRF.5

FRF.11 Voice Over Frame Relay Implementation Agreement (Frame Relay Forum Technical Committee) [21]	■ Addresses the requirements for the transport of compressed voice within the payload of a frame relay frame ■ Supports a diverse set of voice compression algorithms ■ Allows for effective utilization of low bit rate frame relay connections ■ Allows for multiplexing of up to 255 subchannels on a single frame relay DLCI ■ Supports multiple voice payloads on the same or different subframes within a single frame ■ Supports data subchannels on a multiplexed frame relay DLCI
FRF.11 Annex A	Dialed Digit Transfer Syntax
FRF.11 Annex B	Signaling Bit Transfer Syntax
FRF.11 Annex C	Data Transfer Syntax (e.g., RFC 1490 and FRF. 3.1 packets)
FRF.11 Annex D	Fax Relay Transfer Syntax
FRF.11 Annex E	CS-ACELP Transfer Syntax
FRF.11 Annex F	Generic PCM/ADPCM Voice Transfer Syntax
FRF.11 Annex G	G.727 Discard Eligible EADPCM Voice Transfer Syntax
FRF.11 Annex H	G.728 LD-CELP Transfer Syntax
FRF.11 Annex I	G.723.1 MP-MLQ Dual Rate Speech Coder

FRF.8 Frame Relay/ATM PVC Service Interworking Implementation Agreement [20]	Service Interworking applies when a frame relay service user interworks with an ATM service user. FRF.8 is an implementation agreement for *permanent virtual connection* (PVC) service interworking between frame relay and ATM technologies, where the B-ISDN service user employs the B-ISDN Class C AAL5-based message mode via an ATM UNI interface. The ATM service user performs no frame relaying service-specific functions, and the frame relaying service user performs no ATM service-specific functions. All interworking is performed by the *interworking function* (IWF).
FRF.5 Frame Relay/ATM PVC Network Interworking Implementation Agreement [19]	An implementation agreement for PVC network interworking between frame relay and ATM technologies. An IWF provides all mapping and encapsulation functions necessary to ensure that the service provided to the FR terminal equipment is unchanged by the presence of ATM transport.

Appendix

B

Talkspurt Probability Tables

Reducing the Traffic Descriptors for Silence Removal

Table B-1 summarizes the results of Tables B-2 through B-11. It shows how many sources to provision for when reducing PCR and MBS traffic descriptors to accommodate silence suppression (according to the number of voice sources being combined in a single VCC and the selected CLR target).

Guide to the Following Tables

The following tables show the probability that *j* or more sources (column) out of *n* sources (row) will continue talking beyond 1.34 sec, that is, an average-length talkspurt.

TABLE B-1 Summary of Number of Sources to Provision Bandwidth for When Reducing the Traffic Descriptors PCR and MBS to Account for Silence Suppression

Number of Sources in One VCC	Nrt-VBR Class C	Nrt-VBR Class B	Nrt-VBR Class A	Rt-VBR Class A Rt-VBR Class B
	CLR 1.0 e-5	CLR 1.0 e-6	CLR 1.0 e-7	CLR 1.0 e-9
30	22	24	25	27
60	38	40	42	45
90	53	55	57	61
120	66	69	72	76

Scientific notation is used (for example, 9.12 e-04 = 0.000912), which is an equivalent risk to meet a CLR target of 1.0 e-03.

Example For 60 sources, the probability that 40 or more will continue talking at the same time for longer than 1.34 sec is 1.72 e-06. This is sufficient to meet a CLR target of 1.0 e-5. Hence if sufficient bandwidth is provisioned for 39 sources, the risk that the required bandwidth will exceed 39 sources (40 and above) is 1.0 e-5.

Where there is more than one result with the same exponent, it is recommended that you provision the lowest number of sources.

In summary, find the first occurrence in a row of an exponent one higher than the CLR target you are seeking. Reduce the number of sources in the column heading above that value by one, and provision this number of sources.

TABLE B-2 Risk that ≥ j Sources Will Continue Talking Beyond 1.34 sec Out of n Sources (Row) Combined in a Single VCC, Where j = 7 to 14 and n = 7 to 30

j (column) n (row)	≥7	≥8	≥9	≥10	≥11	≥12	≥13	≥14
7	**9.12e-04**							
8	4.37e-03	**3.35e-04**						
9	1.20e-02	1.82e-03	**1.23e-04**					
10	2.49e-02	5.57e-03	**7.47e-04**	**4.54e-05**				
11	4.33e-02	1.27e-02	2.52e-03	**3.04e-04**	**1.67e-05**			
12	6.65e-02	2.39e-02	6.26e-03	1.12e-03	**1.22e-04**	**6.14e-06**		
13	9.33e-02	3.96e-02	1.28e-02	3.01e-03	**4.89e-04**	**4.89e-05**	**2.26e-06**	
14	1.22e-01	5.94e-02	2.26e-02	6.60e-03	1.42e-03	**2.11e-04**	**1.94e-05**	**8.32e-07**
15	1.52e-01	8.26e-02	3.61e-02	1.25e-02	3.32e-03	**6.55e-04**	**8.99e-05**	**7.66e-06**
16	1.82e-01	1.08e-01	5.32e-02	2.12e-02	6.70e-03	1.64e-03	**2.98e-04**	**3.79e-05**
17	2.10e-01	1.35e-01	7.35e-02	3.30e-02	1.20e-02	3.50e-03	**7.90e-04**	**1.33e-04**
18	2.35e-01	1.63e-01	9.62e-02	4.79e-02	1.97e-02	6.64e-03	1.79e-03	**3.75e-04**
19	2.58e-01	1.89e-01	1.21e-01	6.57e-02	3.01e-02	1.15e-02	3.57e-03	**8.94e-04**
20	2.78e-01	2.15e-01	1.46e-01	8.59e-02	4.32e-02	1.83e-02	6.47e-03	1.88e-03
21	2.96e-01	2.38e-01	1.71e-01	1.08e-01	5.89e-02	2.75e-02	1.08e-02	3.57e-03
22	3.10e-01	2.59e-01	1.96e-01	1.31e-01	7.69e-02	3.90e-02	1.69e-02	6.24e-03
23	3.22e-01	2.78e-01	2.19e-01	1.55e-01	9.69e-02	5.30e-02	2.51e-02	1.02e-02
24	3.32e-01	2.94e-01	2.41e-01	1.79e-01	1.18e-01	6.91e-02	3.53e-02	1.57e-02
25	3.40e-01	3.08e-01	2.61e-01	2.02e-01	1.40e-01	8.72e-02	4.78e-02	2.29e-02
26	3.46e-01	3.20e-01	2.78e-01	2.23e-01	1.63e-01	1.07e-01	6.23e-02	3.20e-02
27	3.51e-01	3.30e-01	2.93e-01	2.43e-01	1.85e-01	1.27e-01	7.87e-02	4.32e-02
28	3.55e-01	3.38e-01	3.07e-01	2.62e-01	2.07e-01	1.49e-01	9.66e-02	5.62e-02
29	3.58e-01	3.44e-01	3.18e-01	2.78e-01	2.27e-01	1.70e-01	1.16e-01	7.11e-02
30	3.61e-01	3.49e-01	3.28e-01	2.93e-01	2.46e-01	1.91e-01	1.36e-01	8.75e-02

TABLE B-3 Risk that ≥ j Sources Will Continue Talking Beyond 1.34 sec Out of n Sources (Row) Combined in a Single VCC, Where j = 15 to 22 and n = 15 to 43

j (column) n (row)	≥15	≥16	≥17	≥18	≥19	≥20	≥21	≥22
15	**3.06e-07**							
16	**3.01e-06**	**1.13e-07**						
17	**1.58e-05**	**1.18e-06**	**4.14e-08**					
18	**5.91e-05**	**6.58e-06**	**4.60e-07**	**1.52e-08**				
19	**1.75e-04**	**2.59e-05**	**2.71e-06**	**1.79e-07**	**5.60e-09**			
20	**4.40e-04**	**8.09e-05**	**1.12e-05**	**1.11e-06**	**6.94e-08**	**2.06e-09**		
21	**9.69e-04**	**2.13e-04**	**3.69e-05**	**4.84e-06**	**4.52e-07**	**2.68e-08**	**7.58e-10**	
22	1.93e-03	**4.91e-04**	**1.02e-04**	**1.66e-05**	**2.07e-06**	**1.83e-07**	**1.03e-08**	**2.79e-10**
23	3.51e-03	1.02e-03	**2.45e-04**	**4.79e-05**	**7.42e-06**	**8.76e-07**	**7.40e-08**	**3.98e-09**
24	5.96e-03	1.94e-03	**5.30e-04**	**1.20e-04**	**2.23e-05**	**3.28e-06**	**3.69e-07**	**2.97e-08**
25	9.53e-03	3.42e-03	1.05e-03	**2.71e-04**	**5.84e-05**	**1.03e-05**	**1.44e-06**	**1.55e-07**
26	1.44e-02	5.67e-03	1.92e-03	**5.56e-04**	**1.37e-04**	**2.80e-05**	**4.69e-06**	**6.28e-07**
27	2.09e-02	8.90e-03	3.30e-03	1.06e-03	**2.91e-04**	**6.79e-05**	**1.33e-05**	**2.12e-06**
28	2.91e-02	1.33e-02	5.36e-03	1.88e-03	**5.73e-04**	**1.50e-04**	**3.34e-05**	**6.22e-06**
29	3.91e-02	1.91e-02	8.29e-03	3.16e-03	1.05e-03	**3.06e-04**	**7.63e-05**	**1.62e-05**
30	5.09e-02	2.65e-02	1.23e-02	5.05e-03	1.83e-03	**5.81e-04**	**1.61e-04**	**3.83e-05**
31	6.43e-02	3.54e-02	1.75e-02	7.71e-03	3.01e-03	1.04e-03	**3.15e-04**	**8.33e-05**
32	7.94e-02	4.61e-02	2.41e-02	1.13e-02	4.74e-03	1.77e-03	**5.82e-04**	**1.69e-04**
33	9.58e-02	5.83e-02	3.22e-02	1.60e-02	7.16e-03	2.86e-03	1.02e-03	**3.21e-04**
34	1.13e-01	7.21e-02	4.18e-02	2.20e-02	1.04e-02	4.44e-03	1.70e-03	**5.77e-04**
35	1.31e-01	8.72e-02	5.30e-02	2.93e-02	1.47e-02	6.64e-03	2.70e-03	**9.88e-04**
36	1.50e-01	1.03e-01	6.56e-02	3.80e-02	2.00e-02	9.59e-03	4.15e-03	1.62e-03
37	1.69e-01	1.21e-01	7.95e-02	4.81e-02	2.66e-02	1.34e-02	6.15e-03	2.55e-03
38	1.87e-01	1.38e-01	9.46e-02	5.97e-02	3.45e-02	1.83e-02	8.83e-03	3.88e-03
39	2.05e-01	1.56e-01	1.11e-01	7.25e-02	4.38e-02	2.43e-02	1.23e-02	5.70e-03
40	2.23e-01	1.74e-01	1.27e-01	8.66e-02	5.44e-02	3.14e-02	1.67e-02	8.13e-03
41	2.39e-01	1.92e-01	1.45e-01	1.02e-01	6.62e-02	3.99e-02	2.21e-02	1.13e-02
42	2.55e-01	2.09e-01	1.62e-01	1.17e-01	7.92e-02	4.96e-02	2.87e-02	1.53e-02
43	2.69e-01	2.26e-01	1.80e-01	1.34e-01	9.33e-02	6.05e-02	3.63e-02	2.02e-02

TABLE B-4 Risk that ≥ j Sources Will Continue Talking Beyond 1.34 sec Out of n Sources (Row) Combined in a Single VCC, Where j = 23 to 30 and n = 23 to 54

j (column) n (row)	≥23	≥24	≥25	≥26	≥27	≥28	≥29	≥30
23	**1.03e-10**							
24	**1.53e-09**	**3.77e-11**						
25	**1.19e-08**	**5.87e-10**	**1.39e-11**					
26	**6.44e-08**	**4.75e-09**	**2.25e-10**	5.11e-12				
27	**2.72e-07**	**2.67e-08**	**1.89e-09**	**8.58e-11**	1.88e-12			
28	**9.53e-07**	**1.17e-07**	**1.10e-08**	**7.49e-10**	**3.28e-11**	6.91e-13		
29	**2.89e-06**	**4.24e-07**	**4.99e-08**	**4.52e-09**	**2.96e-10**	**1.25e-11**	2.54e-13	
30	**7.79e-06**	**1.33e-06**	**1.88e-07**	**2.12e-08**	**1.85e-09**	**1.17e-10**	4.76e-12	9.36e-14
31	**1.90e-05**	**3.71e-06**	**6.09e-07**	**8.25e-08**	**8.98e-09**	**7.55e-10**	**4.60e-11**	1.81e-12
32	**4.27e-05**	**9.34e-06**	**1.75e-06**	**2.76e-07**	**3.60e-08**	**3.78e-09**	**3.07e-10**	**1.81e-11**
33	**8.90e-05**	**2.16e-05**	**4.54e-06**	**8.18e-07**	**1.24e-07**	**1.56e-08**	**1.58e-09**	**1.24e-10**
34	**1.74e-04**	**4.64e-05**	**1.08e-05**	**2.19e-06**	**3.79e-07**	**5.56e-08**	**6.75e-09**	**6.62e-10**
35	**3.22e-04**	**9.35e-05**	**2.39e-05**	**5.36e-06**	**1.04e-06**	**1.75e-07**	**2.47e-08**	**2.90e-09**
36	**5.67e-04**	**1.78e-04**	**4.95e-05**	**1.22e-05**	**2.63e-06**	**4.95e-07**	**7.99e-08**	**1.09e-08**
37	**9.55e-04**	**3.21e-04**	**9.67e-05**	**2.59e-05**	**6.15e-06**	**1.28e-06**	**2.33e-07**	**3.63e-08**
38	1.54e-03	**5.54e-04**	**1.79e-04**	**5.19e-05**	**1.34e-05**	**3.07e-06**	**6.18e-07**	**1.08e-07**
39	2.40e-03	**9.18e-04**	**3.17e-04**	**9.88e-05**	**2.76e-05**	**6.88e-06**	**1.52e-06**	**2.96e-07**
40	3.61e-03	1.46e-03	**5.38e-04**	**1.79e-04**	**5.38e-05**	**1.45e-05**	**3.49e-06**	**7.47e-07**
41	5.28e-03	2.25e-03	**8.78e-04**	**3.11e-04**	**9.99e-05**	**2.89e-05**	**7.54e-06**	**1.76e-06**
42	7.49e-03	3.37e-03	1.38e-03	**5.20e-04**	**1.78e-04**	**5.50e-05**	**1.54e-05**	**3.88e-06**
43	1.04e-02	4.88e-03	2.11e-03	**8.38e-04**	**3.03e-04**	**1.00e-04**	**3.00e-05**	**8.13e-06**
44	1.40e-02	6.89e-03	3.13e-03	1.31e-03	**5.00e-04**	**1.75e-04**	**5.58e-05**	**1.62e-05**
45	1.84e-02	9.50e-03	4.52e-03	1.98e-03	**7.97e-04**	**2.95e-04**	**9.96e-05**	**3.07e-05**
46	2.39e-02	1.28e-02	6.35e-03	2.91e-03	1.23e-03	**4.79e-04**	**1.71e-04**	**5.61e-05**
47	3.03e-02	1.69e-02	8.72e-03	4.18e-03	1.85e-03	**7.56e-04**	**2.85e-04**	**9.85e-05**
48	3.77e-02	2.18e-02	1.17e-02	5.85e-03	2.71e-03	1.16e-03	**4.58e-04**	**1.67e-04**
49	4.63e-02	2.77e-02	1.54e-02	8.01e-03	3.86e-03	1.73e-03	**7.16e-04**	**2.74e-04**
50	5.58e-02	3.45e-02	1.99e-02	1.07e-02	5.39e-03	2.51e-03	1.09e-03	**4.37e-04**
51	6.64e-02	4.24e-02	2.53e-02	1.41e-02	7.35e-03	3.57e-03	1.61e-03	**6.76e-04**
52	7.80e-02	5.12e-02	3.16e-02	1.82e-02	9.84e-03	4.96e-03	2.33e-03	1.02e-03
53	9.04e-02	6.11e-02	3.88e-02	2.31e-02	1.29e-02	6.76e-03	3.30e-03	1.50e-03
54	1.04e-01	7.19e-02	4.70e-02	2.89e-02	1.67e-02	9.03e-03	4.57e-03	2.16e-03

TABLE B-5 Risk that ≥ j Sources Will Continue Talking Beyond 1.34 sec Out of n Sources (Row) Combined in a Single VCC, Where j = 31 to 38 and n = 31 to 72

j (column) n (row)	≥31	≥32	≥33	≥34	≥35	≥36	≥37	≥38
31	3.44e-14							
32	6.87e-13	1.27e-14						
33	7.08e-12	2.61e-13	4.66e-15					
34	**5.02e-11**	2.77e-12	9.89e-14	1.71e-15				
35	**2.75e-10**	**2.02e-11**	1.08e-12	3.75e-14	6.30e-16			
36	**1.24e-09**	**1.14e-10**	8.12e-12	4.22e-13	1.42e-14	2.32e-16		
37	**4.81e-09**	**5.29e-10**	**4.71e-11**	3.25e-12	1.64e-13	5.36e-15	8.53e-17	
38	**1.64e-08**	**2.10e-09**	**2.24e-10**	**1.94e-11**	1.30e-12	6.37e-14	2.03e-15	3.14e-17
39	**5.03e-08**	**7.36e-09**	**9.15e-10**	**9.48e-11**	7.95e-12	5.19e-13	2.47e-14	7.66e-16
40	1.41e-07	**2.31e-08**	**3.29e-09**	**3.97e-10**	**3.99e-11**	3.25e-12	2.07e-13	9.58e-15
41	**3.64e-07**	**6.64e-08**	**1.06e-08**	**1.46e-09**	**1.71e-10**	**1.67e-11**	1.33e-12	8.20e-14
42	**8.76e-07**	**1.76e-07**	**3.11e-08**	**4.82e-09**	**6.45e-10**	**7.35e-11**	6.99e-12	5.40e-13
43	**1.98e-06**	**4.33e-07**	**8.43e-08**	**1.45e-08**	**2.18e-09**	**2.84e-10**	**3.15e-11**	2.91e-12
44	**4.24e-06**	**1.00e-06**	**2.13e-07**	**4.02e-08**	**6.71e-09**	**9.82e-10**	**1.24e-10**	**1.34e-11**
45	**8.63e-06**	**2.19e-06**	**5.04e-07**	**1.04e-07**	**1.90e-08**	**3.09e-09**	**4.40e-10**	**5.42e-11**
46	**1.68e-05**	**4.56e-06**	**1.13e-06**	**2.51e-07**	**5.02e-08**	**8.95e-09**	**1.41e-09**	**1.96e-10**
47	**3.12e-05**	**9.05e-06**	**2.39e-06**	**5.73e-07**	**1.24e-07**	**2.41e-08**	**4.19e-09**	**6.44e-10**
48	**5.60e-05**	**1.72e-05**	**4.84e-06**	**1.24e-06**	**2.89e-07**	**6.08e-08**	**1.15e-08**	**1.95e-09**
49	**9.68e-05**	**3.15e-05**	**9.39e-06**	**2.57e-06**	**6.39e-07**	**1.45e-07**	**2.97e-08**	**5.47e-09**
50	**1.62e-04**	**5.55e-05**	**1.75e-05**	**5.08e-06**	**1.35e-06**	**3.27e-07**	**7.20e-08**	**1.44e-08**
51	**2.63e-04**	**9.47e-05**	**3.15e-05**	**9.65e-06**	**2.72e-06**	**7.02e-07**	**1.66e-07**	**3.56e-08**
52	**4.15e-04**	**1.57e-04**	**5.47e-05**	**1.77e-05**	**5.27e-06**	**1.44e-06**	**3.63e-07**	**8.35e-08**
53	**6.38e-04**	**2.52e-04**	**9.22e-05**	**3.13e-05**	**9.84e-06**	**2.85e-06**	**7.61e-07**	**1.86e-07**

j (column) n (row)	≥31	≥32	≥33	≥34	≥35	≥36	≥37	≥38
54	**9.56e-04**	**3.94e-04**	**1.51e-04**	**5.37e-05**	**1.77e-05**	**5.42e-06**	**1.53e-06**	**3.98e-07**
55	1.40e-03	**6.01e-04**	**2.40e-04**	**8.95e-05**	**3.10e-05**	**9.95e-06**	**2.96e-06**	**8.14e-07**
56	2.01e-03	**8.95e-04**	**3.73e-04**	**1.45e-04**	**5.25e-05**	**1.77e-05**	**5.53e-06**	**1.60e-06**
57	2.82e-03	1.30e-03	**5.65e-04**	**2.29e-04**	**8.65e-05**	**3.05e-05**	**1.00e-05**	**3.05e-06**
58	3.88e-03	1.86e-03	**8.37e-04**	**3.52e-04**	**1.39e-04**	**5.11e-05**	**1.75e-05**	**5.61e-06**
59	5.25e-03	2.60e-03	1.21e-03	**5.31e-04**	**2.17e-04**	**8.34e-05**	**2.99e-05**	**1.00e-05**
60	6.98e-03	3.58e-03	1.73e-03	**7.82e-04**	**3.33e-04**	**1.33e-04**	**4.96e-05**	**1.73e-05**
61	9.13e-03	4.83e-03	2.41e-03	1.13e-03	**4.98e-04**	**2.06e-04**	**8.01e-05**	**2.92e-05**
62	1.18e-02	6.41e-03	3.30e-03	1.60e-03	**7.30e-04**	**3.14e-04**	**1.27e-04**	**4.79e-05**
63	1.49e-02	8.38e-03	4.44e-03	2.22e-03	1.05e-03	**4.67e-04**	**1.95e-04**	**7.68e-05**
64	1.87e-02	1.08e-02	5.89e-03	3.04e-03	1.48e-03	**6.81e-04**	**2.95e-04**	**1.20e-04**
65	2.31e-02	1.37e-02	7.70e-03	4.09e-03	2.06e-03	**9.76e-04**	**4.37e-04**	**1.85e-04**
66	2.82e-02	1.72e-02	9.91e-03	5.42e-03	2.80e-03	1.37e-03	**6.35e-04**	**2.78e-04**
67	3.40e-02	2.12e-02	1.26e-02	7.07e-03	3.76e-03	1.90e-03	**9.07e-04**	**4.09e-04**
68	4.06e-02	2.59e-02	1.58e-02	9.09e-03	4.98e-03	2.59e-03	1.27e-03	**5.92e-04**
69	4.79e-02	3.13e-02	1.95e-02	1.15e-02	6.49e-03	3.47e-03	1.76e-03	**8.42e-04**
70	5.60e-02	3.74e-02	2.39e-02	1.45e-02	8.35e-03	4.58e-03	2.38e-03	1.18e-03
71	6.48e-02	4.43e-02	2.88e-02	1.79e-02	1.06e-02	5.97e-03	3.19e-03	1.62e-03
72	7.42e-02	5.18e-02	3.45e-02	2.19e-02	1.33e-02	7.67e-03	4.21e-03	2.20e-03

TABLE B-6 Risk that ≥ j Sources Will Continue Talking Beyond 1.34 sec Out of n Sources (Row) Combined in a Single VCC, Where j = 39 to 46 and n = 45 to 88

j (column) n (row)	≥39	≥40	≥41	≥42	≥43	≥44	≥45	≥46
45	5.70e-12	5.01e-13	3.59e-14	2.01e-15	8.21e-17	2.19e-18	2.86e-20	
46	**2.35e-11**	2.41e-12	2.07e-13	1.45e-14	7.90e-16	3.16e-17	8.25e-19	1.05e-20
47	**8.70e-11**	**1.02e-11**	1.02e-12	8.54e-14	5.82e-15	3.10e-16	1.21e-17	3.10e-19
48	**2.92e-10**	**3.84e-11**	4.39e-12	4.29e-13	3.51e-14	2.34e-15	1.22e-16	4.66e-18
49	**9.01e-10**	**1.32e-10**	**1.69e-11**	1.89e-12	1.80e-13	1.44e-14	9.37e-16	4.78e-17
50	**2.58e-09**	**4.15e-10**	**5.92e-11**	7.42e-12	8.08e-13	7.53e-14	5.88e-15	3.75e-16
51	**6.92e-09**	**1.21e-09**	**1.90e-10**	**2.64e-11**	3.24e-12	3.45e-13	3.14e-14	2.40e-15
52	**1.75e-08**	**3.31e-09**	**5.66e-10**	**8.66e-11**	**1.18e-11**	1.41e-12	1.47e-13	1.31e-14
53	**4.17e-08**	**8.52e-09**	**1.58e-09**	**2.63e-10**	**3.93e-11**	5.22e-12	6.11e-13	6.22e-14
54	**9.49e-08**	**2.07e-08**	**4.13e-09**	**7.46e-10**	**1.22e-10**	**1.78e-11**	2.31e-12	2.64e-13
55	**2.06e-07**	**4.80e-08**	**1.02e-08**	**1.99e-09**	**3.51e-10**	**5.59e-11**	7.99e-12	1.02e-12
56	**4.30e-07**	**1.06e-07**	**2.41e-08**	**5.02e-019**	**9.54e-10**	**1.65e-10**	**2.56e-11**	3.58e-12
57	**8.62e-07**	**2.25e-07**	**5.44e-08**	**1.21e-08**	**2.45e-09**	**4.55e-10**	**7.67e-11**	**1.17e-11**
58	**1.67e-06**	**4.60e-07**	**1.17e-07**	**2.76e-08**	**5.99e-09**	**1.19e-09**	**2.16e-10**	**3.56e-11**
59	**3.12e-06**	**9.04e-07**	**2.43e-07**	**6.06e-08**	**1.39e-08**	**2.95e-09**	**5.74e-10**	**1.02e-10**
60	**5.65e-06**	**1.72e-06**	**4.86e-07**	**1.28e-07**	**3.11e-08**	**7.00e-09**	**1.45e-09**	**2.76e-10**
61	**9.94e-06**	**3.16e-06**	**9.39e-07**	**2.60e-07**	**6.67e-08**	**1.59e-08**	**3.49e-09**	**7.08e-10**
62	**1.70e-05**	**5.66e-06**	**1.76e-06**	**5.10e-07**	**1.38e-07**	**3.46e-08**	**8.04e-09**	**1.73e-09**
63	**2.84e-05**	**9.84e-06**	**3.19e-06**	**9.69e-07**	**2.75e-07**	**7.25e-08**	**1.78e-08**	**4.05e-09**
64	**4.62e-05**	**1.67e-05**	**5.64e-06**	**1.79e-06**	**5.30e-07**	**1.47e-07**	**3.79e-08**	**9.11e-09**
65	**7.35e-05**	**2.75e-05**	**9.69e-06**	**3.20e-06**	**9.92e-07**	**2.88e-07**	**7.80e-08**	**1.97e-08**
66	**1.14e-04**	**4.44e-05**	**1.63e-05**	**5.59e-06**	**1.81e-06**	**5.47e-07**	**1.55e-07**	**4.11e-08**
67	**1.74e-04**	**7.02e-05**	**2.66e-05**	**9.51e-06**	**3.20e-06**	**1.01e-06**	**2.99e-07**	**8.31e-08**

j (column) n (row)	≥39	≥40	≥41	≥42	≥43	≥44	≥45	≥46
68	**2.61e-04**	**1.09e-04**	**4.27e-05**	**1.58e-05**	**5.52e-06**	**1.81e-06**	**5.61e-07**	**1.63e-07**
69	**3.83e-04**	**1.65e-04**	**6.69e-05**	**2.57e-05**	**9.31e-06**	**3.18e-06**	**1.02e-06**	**3.09e-07**
70	**5.52e-04**	**2.45e-04**	**1.03e-04**	**4.08e-05**	**1.53e-05**	**5.43e-06**	**1.82e-06**	**5.71e-07**
71	**7.82e-04**	**3.58e-04**	**1.55e-04**	**6.36e-05**	**2.47e-05**	**9.07e-06**	**3.15e-06**	**1.03e-06**
72	1.09e-03	**5.14e-04**	**2.30e-04**	**9.73e-05**	**3.90e-05**	**1.48e-05**	**5.33e-06**	**1.81e-06**
73	1.50e-03	**7.26e-04**	**3.34e-04**	**1.46e-04**	**6.05e-05**	**2.37e-05**	**8.82e-06**	**3.10e-06**
74	2.03e-03	1.01e-03	**4.78e-04**	**2.15e-04**	**9.19e-05**	**3.72e-05**	**1.43e-05**	**5.21e-06**
75	2.71e-03	1.39e-03	**6.74e-04**	**3.12e-04**	**1.37e-04**	**5.74e-05**	**2.27e-05**	**8.55e-06**
76	3.57e-03	1.87e-03	**9.36e-04**	**4.45e-04**	**2.02e-04**	**8.68e-05**	**3.55e-05**	**1.38e-05**
77	4.64e-03	2.50e-03	1.28e-03	**6.26e-04**	**2.91e-04**	**1.29e-04**	**5.43e-05**	**2.18e-05**
78	5.96e-03	3.29e-03	1.73e-03	**8.66e-04**	**4.14e-04**	**1.89e-04**	**8.18e-05**	**3.37e-05**
79	7.57e-03	4.27e-03	2.30e-03	1.18e-03	**5.81e-04**	**2.72e-04**	**1.21e-04**	**5.14e-05**
80	9.50e-03	5.48e-03	3.02e-03	1.59e-03	**8.02e-04**	**3.85e-04**	**1.76e-04**	**7.71e-05**
81	1.18e-02	6.96e-03	3.93e-03	2.12e-03	1.09e-03	**5.39e-04**	**2.53e-04**	**1.14e-04**
82	1.45e-02	8.74e-03	5.04e-03	2.79e-03	1.47e-03	**7.43e-04**	**3.58e-04**	**1.65e-04**
83	1.76e-02	1.09e-02	6.40e-03	3.62e-03	1.96e-03	1.01e-03	**5.00e-04**	**2.36e-04**
84	2.13e-02	1.34e-02	8.04e-03	4.64e-03	2.57e-03	1.36e-03	**6.88e-04**	**3.33e-04**
85	2.54e-02	1.63e-02	1.00e-02	5.89e-03	3.33e-03	1.80e-03	**9.34e-04**	**4.64e-04**
86	3.01e-02	1.96e-02	1.23e-02	7.40e-03	4.27e-03	2.36e-03	1.25e-03	**6.37e-04**
87	3.53e-02	2.35e-02	1.50e-02	9.21e-03	5.42e-03	3.07e-03	1.66e-03	**8.64e-04**
88	4.11e-02	2.78e-02	1.81e-02	1.13e-02	6.82e-03	3.93e-03	2.18e-03	1.16e-03

TABLE B-7 Risk that ≥ j Sources Will Continue Talking Beyond 1.34 sec Out of n Sources (Row) Combined in a Single VCC, Where j = 47 to 54 and n = 58 to 107

j (column) n (row)	≥47	≥48	≥49	≥50	≥51	≥52	≥53	≥54
58	**5.31e-12**	**7.12e-13**	**8.49e-14**	**8.94e-15**	**8.21e-16**	**6.47e-17**	**4.29e-18**	**2.33e-19**
59	1.65e-11	**2.40e-12**	**3.16e-13**	**3.69e-14**	**3.81e-15**	**3.43e-16**	**2.65e-17**	**1.73e-18**
60	4.79e-11	**7.58e-12**	**1.08e-12**	**1.39e-13**	**1.60e-14**	**1.62e-15**	**1.43e-16**	**1.08e-17**
61	1.32e-10	2.24e-11	**3.47e-12**	**4.87e-13**	**6.14e-14**	**6.90e-15**	**6.86e-16**	**5.94e-17**
62	3.44e-10	6.26e-11	1.04e-11	**1.59e-12**	**2.18e-13**	**2.69e-14**	**2.97e-15**	**2.90e-16**
63	8.54e-10	1.66e-10	2.96e-11	**4.84e-12**	**7.21e-13**	**9.72e-14**	**1.18e-14**	**1.28e-15**
64	2.03e-09	4.19e-10	7.98e-11	1.40e-11	**2.24e-12**	**3.27e-13**	**4.32e-14**	**5.15e-15**
65	4.63e-09	1.01e-09	2.05e-10	3.82e-11	**6.55e-12**	**1.03e-12**	**1.47e-13**	**1.92e-14**
66	1.02e-08	2.34e-09	5.02e-10	9.94e-11	1.82e-11	**3.06e-12**	**4.72e-13**	**6.64e-14**
67	2.16e-08	5.23e-09	1.18e-09	2.47e-10	4.81e-11	**8.63e-12**	**1.42e-12**	**2.16e-13**
68	4.42e-08	1.12e-08	2.67e-09	5.90e-10	1.21e-10	2.31e-11	**4.07e-12**	**6.60e-13**
69	8.78e-08	2.34e-08	5.82e-09	1.35e-09	2.94e-10	5.93e-11	1.11e-11	**1.92e-12**
70	1.69e-07	4.71e-08	1.23e-08	3.00e-09	6.84e-10	1.46e-10	2.88e-11	**5.29e-12**
71	3.17e-07	9.20e-08	2.51e-08	6.41e-09	1.54e-09	3.44e-10	7.18e-11	1.39e-11
72	5.79e-07	1.75e-07	4.97e-08	1.33e-08	3.33e-09	7.82e-10	1.72e-10	3.52e-11
73	1.03e-06	3.23e-07	9.57e-08	2.67e-08	6.99e-09	1.72e-09	3.96e-10	8.55e-11
74	1.79e-06	5.84e-07	1.80e-07	5.21e-08	1.42e-08	3.66e-09	8.83e-10	2.00e-10
75	3.05e-06	1.03e-06	3.28e-07	9.90e-08	2.81e-08	7.55e-09	1.90e-09	4.51e-10
76	5.07e-06	1.77e-06	5.86e-07	1.83e-07	5.42e-08	1.51e-08	3.98e-09	9.85e-1
77	8.28e-06	2.99e-06	1.02e-06	3.31e-07	1.02e-07	2.95e-08	8.08e-09	2.09e-09
78	1.32e-05	4.93e-06	1.74e-06	5.86e-07	1.86e-07	5.61e-08	1.60e-08	4.29e-09
79	2.08e-05	7.99e-06	2.92e-06	1.01e-06	3.33e-07	1.04e-07	3.07e-08	8.58e-09
80	3.21e-05	1.27e-05	4.78e-06	1.71e-06	5.83e-07	1.88e-07	5.76e-08	1.67e-08
81	4.86e-05	1.98e-05	7.69e-06	2.84e-06	9.99e-07	3.33e-07	1.06e-07	3.18e-08

j (column) n (row)	≥47	≥48	≥49	≥50	≥51	≥52	≥53	≥54
82	7.25e-05	3.04e-05	1.22e-05	4.63e-06	1.68e-06	5.78e-07	1.89e-07	5.90e-08
83	1.07e-04	4.59e-05	1.89e-05	7.40e-06	2.76e-06	9.82e-07	3.32e-07	1.07e-07
84	1.54e-04	6.82e-05	2.88e-05	1.16e-05	4.47e-06	1.64e-06	5.72e-07	1.90e-07
85	2.20e-04	9.99e-05	4.33e-05	1.79e-05	7.10e-06	2.68e-06	9.63e-07	3.30e-07
86	3.10e-04	1.44e-04	6.41e-05	2.73e-05	1.11e-05	4.30e-06	1.59e-06	5.63e-07
87	4.30e-04	2.05e-04	9.35e-05	4.08e-05	1.70e-05	6.80e-06	2.59e-06	9.42e-07
88	5.90e-04	2.88e-04	1.35e-04	6.02e-05	2.58e-05	1.06e-05	4.14e-06	1.55e-06
89	7.99e-04	3.99e-04	1.91e-04	8.76e-05	3.85e-05	1.62e-05	6.50e-06	2.50e-06
90	**1.07e-03**	5.46e-04	2.67e-04	1.26e-04	5.65e-05	2.44e-05	1.01e-05	3.97e-06
91	**1.41e-03**	7.38e-04	3.70e-04	1.78e-04	8.19e-05	3.62e-05	1.53e-05	6.21e-06
92	**1.85e-03**	9.87e-04	5.05e-04	2.48e-04	1.17e-04	5.30e-05	2.30e-05	9.56e-06
93	**2.40e-03**	**1.31e-03**	6.83e-04	3.43e-04	1.65e-04	7.66e-05	3.40e-05	1.45e-05
94	**3.07e-03**	**1.71e-03**	912e-04	4.68e-04	2.31e-04	1.09e-04	4.97e-05	2.17e-05
95	**3.90e-03**	**2.21e-03**	**1.20e-03**	6.31e-04	3.18e-04	1.54e-04	7.16e-05	3.20e-05
96	**4.91e-03**	**2.83e-03**	**1.57e-03**	8.42e-04	4.33e-04	2.14e-04	1.02e-04	4.66e-05
97	**6.11e-03**	**3.60e-03**	**2.04e-03**	**1.11e-03**	5.84e-04	2.95e-04	1.43e-04	6.69e-05
98	**7.55e-03**	**4.52e-03**	**2.61e-03**	**1.45e-03**	7.78e-04	4.01e-04	1.99e-04	9.50e-05
99	**9.24e-03**	**5.63e-03**	**3.31e-03**	**1.88e-03**	**1.03e-03**	5.40e-04	2.73e-04	1.33e-04
100	**1.12e-02**	**6.96e-03**	**4.17e-03**	**2.41e-03**	**1.34e-03**	7.18e-04	3.71e-04	1.85e-04
101	**1.35e-02**	**8.53e-03**	**5.19e-03**	**3.05e-03**	**1.73e-03**	9.47e-04	4.99e-04	2.53e-04
102	**1.61e-02**	**1.04e-02**	**6.42e-03**	**3.84e-03**	**2.22e-03**	**1.24e-03**	6.64e-04	3.44e-04
103	**1.91e-02**	**1.25e-02**	**7.87e-03**	**4.79e-03**	**2.82e-03**	**1.60e-03**	8.74e-04	4.61e-04
104	**2.25e-02**	**1.49e-02**	**9.57e-03**	**5.92e-03**	**3.54e-03**	**2.05e-03**	**1.14e-03**	6.13e-04
105	**2.64e-02**	**1.77e-02**	**1.15e-02**	**7.26e-03**	**4.42e-03**	**2.60e-03**	**1.47e-03**	8.07e-04
106	**3.06e-02**	**2.09e-02**	**1.38e-02**	**8.84e-03**	**5.46e-03**	**3.27e-03**	**1.89e-03**	**1.05e-03**
107	**3.53e-02**	**2.45e-02**	**1.64e-02**	**1.07e-02**	**6.71e-03**	**4.08e-03**	**2.39e-03**	**1.36e-03**

TABLE B-8 Risk That ≥ j Sources Will Continue Talking Beyond 1.34 sec Out of n Sources (Row) Combined in a Single VCC, Where j = 55 to 62 and n = 72 to 120

j (column) n (row)	≥55	≥56	≥57	≥58	≥59	≥61	≥61	≥62
72	**6.72e-12**	**1.19e-12**	**1.95e-13**	**2.94e-14**	**4.07e-15**	**5.15e-16**	**5.92e-17**	**6.14e-18**
73	1.72e-11	**3.22e-12**	**5.61e-13**	**9.02e-14**	**1.34e-14**	**1.82e-15**	**2.27e-16**	**2.57e-17**
74	4.23e-11	8.37e-12	1.54e-12	**2.63e-13**	**4.17e-14**	**6.08e-15**	**8.15e-16**	**9.98e-17**
75	1.00e-10	2.09e-11	4.05e-12	**7.33e-13**	**1.23e-13**	**1.92e-14**	**2.75e-15**	**3.63e-16**
76	2.29e-10	5.01e-11	1.02e-11	**1.95e-12**	**3.48e-13**	**5.74e-14**	**8.79e-15**	**1.24e-15**
77	5.08e-10	1.16e-10	2.49e-11	**5.00e-12**	**9.38e-13**	**1.64e-13**	**2.67e-14**	**4.02e-15**
78	1.09e-09	2.60e-10	5.84e-11	1.23e-11	**2.43e-12**	**4.49e-13**	**7.73e-14**	**1.24e-14**
79	2.27e-09	5.65e-10	1.33e-10	2.93e-11	**6.07e-12**	**1.18e-12**	**2.14e-13**	**3.62e-14**
80	4.59e-09	1.19e-09	2.92e-10	6.73e-11	1.46e-11	**2.98e-12**	**5.69e-13**	**1.02e-13**
81	9.05e-09	2.44e-09	6.22e-10	1.50e-10	3.40e-11	**7.26e-12**	**1.46e-12**	**2.74e-13**
82	1.74e-08	4.87e-09	1.29e-09	3.24e-10	7.66e-11	1.71e-11	**3.59e-12**	**7.08e-13**
83	3.27e-08	9.49e-09	2.61e-09	6.80e-10	1.67e-10	3.90e-11	**8.55e-12**	**1.77e-12**
84	6.00e-08	1.80e-08	5.14e-09	1.39e-09	3.56e-10	8.63e-11	1.97e-11	**4.26e-12**
85	1.08e-07	3.35e-08	9.88e-09	2.77e-09	7.36e-10	1.85e-10	4.42e-11	**9.96e-12**
86	1.90e-07	6.08e-08	1.86e-08	5.39e-09	1.48e-09	3.88e-10	9.62e-11	2.26e-11
87	3.27e-07	1.08e-07	3.41e-08	1.02e-08	2.92e-09	7.91e-10	2.04e-10	4.96e-11
88	5.53e-07	1.89e-07	6.14e-08	1.90e-08	5.61e-09	1.57e-09	4.20e-10	1.06e-10
89	9.20e-07	3.23e-07	1.08e-07	3.46e-08	1.05e-08	3.06e-09	8.44e-10	2.22e-10
90	1.50e-06	5.42e-07	1.87e-07	6.17e-08	1.94e-08	5.81e-09	1.66e-09	4.51e-10
91	2.41e-06	8.95e-07	3.18e-07	1.08e-07	3.49e-08	1.08e-08	3.19e-09	8.95e-10
92	3.81e-06	1.45e-06	5.30e-07	1.85e-07	6.18e-08	1.97e-08	5.99e-09	1.74e-09
93	5.93e-06	2.32e-06	8.70e-07	3.12e-07	1.07e-07	3.52e-08	1.10e-08	3.30e-09
94	9.08e-06	3.65e-06	1.40e-06	5.17e-07	1.83e-07	6.17e-08	1.99e-08	6.15e-09
95	1.37e-05	5.65e-06	2.23e-06	8.43e-07	3.06e-07	1.06e-07	3.53e-08	1.12e-08

j (column) n (row)	≥55	≥56	≥57	≥58	≥59	≥61	≥61	≥62
96	**2.04e-05**	**8.62e-06**	**3.49e-06**	**1.35e-06**	**5.03e-07**	**1.80e-07**	**6.13e-08**	**2.01e-08**
97	**3.01e-05**	**1.30e-05**	**5.37e-06**	**2.14e-06**	**8.16e-07**	**2.99e-07**	**1.05e-07**	**3.52e-08**
98	**4.36e-05**	**1.93e-05**	**8.17e-06**	**3.33e-06**	**1.30e-06**	**4.89e-07**	**1.76e-07**	**6.08e-08**
99	**6.25e-05**	**2.82e-05**	**1.22e-05**	**5.11e-06**	**2.05e-06**	**7.88e-07**	**2.91e-07**	**1.03e-07**
100	**8.86e-05**	**4.08e-05**	**1.81e-05**	**7.73e-06**	**3.17e-06**	**1.25e-06**	**4.74e-07**	**1.72e-07**
101	**1.24e-04**	**5.84e-05**	**2.65e-05**	**1.16e-05**	**4.85e-06**	**1.96e-06**	**7.60e-07**	**2.83e-07**
102	**1.72e-04**	**8.25e-05**	**3.82e-05**	**1.70e-05**	**7.32e-06**	**3.02e-06**	**1.20e-06**	**4.59e-07**
103	**2.35e-04**	**1.15e-04**	**5.45e-05**	**2.48e-05**	**1.09e-05**	**4.60e-06**	**1.87e-06**	**7.32e-07**
104	**3.18e-04**	**1.59e-04**	**7.69e-05**	**3.58e-05**	**1.60e-05**	**6.92e-06**	**2.88e-06**	**1.15e-06**
105	**4.27e-04**	**2.18e-04**	**1.07e-04**	**5.09e-05**	**2.33e-05**	**1.03e-05**	**4.36e-06**	**1.79e-06**
106	**5.67e-04**	**2.95e-04**	**1.48e-04**	**7.16e-05**	**3.34e-05**	**1.51e-05**	**6.54e-06**	**2.73e-06**
107	**7.45e-04**	**3.95e-04**	**2.02e-04**	**9.97e-05**	**4.75e-05**	**2.18e-05**	**9.67e-06**	**4.13e-06**
108	**9.71e-04**	**5.24e-04**	**2.73e-04**	**1.37e-04**	**6.67e-05**	**3.13e-05**	**1.41e-05**	**6.17e-06**
109	1.25e-03	**6.88e-04**	**3.65e-04**	**1.87e-04**	**9.26e-05**	**4.43e-05**	**2.04e-05**	**9.10e-06**
110	1.60e-03	**8.96e-04**	**4.84e-04**	**2.53e-04**	**1.27e-04**	**6.21e-05**	**2.92e-05**	**1.33e-05**
111	2.04e-03	1.16e-03	**6.36e-04**	**3.38e-04**	**1.73e-04**	**8.61e-05**	**4.13e-05**	**1.91e-05**
112	2.56e-03	1.48e-03	**8.27e-04**	**4.47e-04**	**2.34e-04**	**1.18e-04**	**5.78e-05**	**2.73e-05**
113	3.20e-03	1.88e-03	1.07e-03	**5.87e-04**	**3.12e-04**	**1.61e-04**	**8.00e-05**	**3.85e-05**
114	3.97e-03	2.37e-03	1.37e-03	**7.64e-04**	**4.13e-04**	**2.17e-04**	**1.10e-04**	**5.38e-05**
115	4.88e-03	2.95e-03	1.73e-03	**9.85e-04**	**5.42e-04**	**2.89e-04**	**1.49e-04**	**7.44e-05**
116	5.95e-03	3.66e-03	2.18e-03	1.26e-03	**7.05e-04**	**3.82e-04**	**2.01e-04**	**1.02e-04**
117	7.21e-03	4.50e-03	2.73e-03	1.60e-03	**9.10e-04**	**5.01e-04**	**2.67e-04**	**1.38e-0**
118	8.68e-03	5.50e-03	3.38e-03	2.01e-03	1.16e-03	**6.51e-04**	**3.53e-04**	**1.86e-04**
119	1.04e-02	6.67e-03	4.16e-03	2.52e-03	1.48e-03	**8.40e-04**	**4.63e-04**	**2.47e-04**
120	1.23e-02	8.03e-03	5.08e-03	3.12e-03	1.86e-03	1.07e-03	**6.02e-04**	**3.27e-04**

TABLE B-9 Risk That ≥ j Sources Will Continue Talking Beyond 1.34 sec Out of n Sources (Row) Combined in a Single VCC, Where j = 63 to 70 and n = 86 to 120

j (column) n (row)	≥63	≥64	≥65	≥66	≥67	≥68	≥69	≥70
86	5.00e-12	1.05e-12	2.06e-13	3.83e-14	6.66e-15	1.09e-15	1.65e-16	2.34e-17
87	**1.15e-11**	2.50e-12	5.16e-13	1.00e-13	1.83e-14	3.14e-15	5.04e-16	7.55e-17
88	**2.55e-11**	5.80e-12	1.25e-12	2.53e-13	4.84e-14	8.71e-15	1.47e-15	2.33e-16
89	**5.52e-11**	**1.30e-11**	2.92e-12	6.18e-13	1.24e-13	2.33e-14	4.14e-15	6.89e-16
90	**1.16e-10**	**2.86e-11**	6.65e-12	1.47e-12	3.06e-13	6.02e-14	1.12e-14	1.96e-15
91	**2.39e-10**	**6.09e-11**	**1.47e-11**	3.37e-12	7.32e-13	1.51e-13	2.92e-14	5.35e-15
92	**4.81e-10**	**1.27e-10**	**3.17e-11**	7.54e-12	1.70e-12	3.65e-13	7.38e-14	1.41e-14
93	**9.43e-10**	**2.57e-10**	**6.66e-11**	**1.64e-11**	3.85e-12	8.57e-13	1.81e-13	3.61e-14
94	**1.81e-09**	**5.09e-10**	**1.37e-10**	**3.49e-11**	8.48e-12	1.96e-12	4.30e-13	8.93e-14
95	**3.41e-09**	**9.88e-10**	**2.74e-10**	**7.23e-11**	**1.82e-11**	4.36e-12	9.92e-13	2.14e-13
96	**6.28e-09**	**1.88e-09**	**5.37e-10**	**1.46e-10**	**3.81e-11**	9.45e-12	2.23e-12	5.01e-13
97	**1.13e-08**	**3.50e-09**	**1.03e-09**	**2.90e-10**	**7.79e-11**	**2.00e-11**	4.88e-12	1.14e-12
98	**2.01e-08**	**6.38e-09**	**1.94e-09**	**5.62e-10**	**1.56e-10**	**4.13e-11**	**1.04e-11**	2.52e-12
99	**3.51e-08**	**1.14e-08**	**3.57e-09**	**1.07e-09**	**3.05e-10**	**8.35e-11**	**2.18e-11**	5.43e-12
100	**6.02e-08**	**2.02e-08**	**6.47e-09**	**1.99e-09**	**5.86e-10**	**1.65e-10**	**4.45e-11**	**1.15e-11**
101	**1.01e-07**	**3.49e-08**	**1.15e-08**	**3.64e-09**	**1.10e-09**	**3.20e-10**	**8.89e-11**	**2.36e-11**
102	**1.68e-07**	**5.94e-08**	**2.01e-08**	**6.53e-09**	**2.03e-09**	**6.08e-10**	**1.74e-10**	**4.76e-11**
103	**2.75e-07**	**9.95e-08**	**3.45e-08**	**1.15e-08**	**3.69e-09**	**1.13e-09**	**3.33e-10**	**9.41e-11**
104	**4.43e-07**	**1.64e-07**	**5.84e-08**	**2.00e-08**	**6.57e-09**	**2.07e-09**	**6.27e-10**	**1.82e-10**
105	**7.04e-07**	**2.67e-07**	**9.73e-08**	**3.41e-08**	**1.15e-08**	**3.73e-09**	**1.16e-09**	**3.46e-10**

j (column) n (row)	≥63	≥64	≥65	≥66	≥67	≥68	≥69	≥70
106	**1.10e-06**	**4.27e-07**	**1.60e-07**	**5.74e-08**	**1.98e-08**	**6.59e-09**	**2.10e-09**	**6.45e-10**
107	**1.70e-06**	**6.75e-07**	**2.58e-07**	**9.50e-08**	**3.36e-08**	**1.15e-08**	**3.75e-09**	**1.18e-09**
108	**2.60e-06**	**1.05e-06**	**4.12e-07**	**1.55e-07**	**5.62e-08**	**1.96e-08**	**6.59e-09**	**2.13e-09**
109	**3.91e-06**	**1.62e-06**	**6.48e-07**	**2.49e-07**	**9.26e-08**	**3.31e-08**	**1.14e-08**	**3.77e-09**
110	**5.82e-06**	**2.46e-06**	**1.01e-06**	**3.96e-07**	**1.50e-07**	**5.50e-08**	**1.94e-08**	**6.57e-09**
111	**8.56e-06**	**3.70e-06**	**1.54e-06**	**6.20e-07**	**2.41e-07**	**9.00e-08**	**3.25e-08**	**1.13e-08**
112	**1.25e-05**	**5.49e-06**	**2.34e-06**	**9.59e-07**	**3.80e-07**	**1.45e-07**	**5.36e-08**	**1.91e-08**
113	**1.79e-05**	**8.05e-06**	**3.49e-06**	**1.47e-06**	**5.93e-07**	**2.32e-07**	**8.74e-08**	**3.18e-08**
114	**2.55e-05**	**1.17e-05**	**5.17e-06**	**2.21e-06**	**9.14e-07**	**3.65e-07**	**1.41e-07**	**5.23e-08**
115	**3.59e-05**	**1.68e-05**	**7.56e-06**	**3.30e-06**	**1.39e-06**	**5.67e-07**	**2.23e-07**	**8.47e-08**
116	**5.00e-05**	**2.38e-05**	**1.09e-05**	**4.87e-06**	**2.09e-06**	**8.70e-07**	**3.50e-07**	**1.36e-07**
117	**6.91e-05**	**3.35e-05**	**1.57e-05**	**7.10e-06**	**3.11e-06**	**1.32e-06**	**5.41e-07**	**2.14e-07**
118	**9.45e-05**	**4.66e-05**	**2.22e-05**	**1.03e-05**	**4.58e-06**	**1.98e-06**	**8.28e-07**	**3.35e-07**
119	**1.28e-04**	**6.42e-05**	**3.12e-05**	**1.47e-05**	**6.67e-06**	**2.94e-06**	**1.25e-06**	**5.16e-07**
120	**1.72e-04**	**8.77e-05**	**4.33e-05**	**2.07e-05**	**9.61e-06**	**4.31e-06**	**1.87e-06**	**7.87e-07**

TABLE B-10 Risk That ≥ j Sources Will Continue Talking Beyond 1.34 sec Out of n Sources (Row) Combined in a Single VCC, Where j = 71 to 78 and n = 101 to 120

j (column) n (row)	≥71	≥72	≥73	≥74	≥75	≥76	≥77	≥78
101	5.99e-12	1.45e-12	3.36e-13	7.39e-14	1.55e-14	3.09e-15	5.84e-16	1.05e-16
102	**1.25e-11**	3.12e-12	7.46e-13	1.70e-13	3.70e-14	7.65e-15	1.50e-15	2.81e-16
103	**2.54e-11**	**6.56e-12**	1.62e-12	3.82e-13	8.60e-14	1.84e-14	3.76e-15	7.31e-16
104	**5.07e-11**	**1.35e-11**	3.44e-12	8.38e-13	1.95e-13	4.33e-14	9.16e-15	1.85e-15
105	**9.90e-11**	**2.72e-11**	7.14e-12	1.79e-12	4.31e-13	9.91e-14	2.17e-14	4.54e-15
106	**1.90e-10**	**5.36e-11**	**1.45e-11**	3.76e-12	9.33e-13	2.21e-13	5.02e-14	1.09e-14
107	**3.57e-10**	**1.04e-10**	**2.89e-11**	7.71e-12	1.97e-12	4.83e-13	1.13e-13	2.53e-14
108	**6.61e-10**	**1.97e-10**	**5.64e-11**	**1.55e-11**	4.08e-12	1.03e-12	2.49e-13	5.76e-14
109	**1.20e-09**	**3.68e-10**	**1.08e-10**	**3.06e-11**	8.28e-12	2.15e-12	5.37e-13	1.28e-13
110	**2.15e-09**	**6.74e-10**	**2.04e-10**	**5.91e-11**	**1.65e-11**	4.41e-12	1.13e-12	2.78e-13
111	**3.77e-09**	**1.22e-09**	**3.77e-10**	**1.12e-10**	**3.22e-11**	8.85e-12	2.34e-12	5.92e-13
112	**6.53e-09**	**2.16e-09**	**6.85e-10**	**2.10e-10**	**6.16e-11**	**1.74e-11**	4.73e-12	1.23e-12
113	**1.11e-08**	**3.77e-09**	**1.23e-09**	**3.85e-10**	**1.16e-10**	**3.37e-11**	9.40e-12	2.52e-12
114	**1.87e-08**	**6.48e-09**	**2.16e-09**	**6.94e-10**	**2.15e-10**	**6.40e-11**	**1.83e-11**	5.05e-12
115	**3.11e-08**	**1.10e-08**	**3.75e-09**	**1.23e-09**	**3.91e-10**	**1.19e-10**	**3.51e-11**	9.94e-12
116	**5.08e-08**	**1.84e-08**	**6.41e-09**	**2.16e-09**	**7.01e-10**	**2.19e-10**	**6.62e-11**	**1.92e-11**
117	**8.20e-08**	**3.03e-08**	**1.08e-08**	**3.72e-09**	**1.24e-09**	**3.97e-10**	**1.23e-10**	**3.65e-11**
118	**1.31e-07**	**4.93e-08**	**1.80e-08**	**6.33e-09**	**2.15e-09**	**7.06e-10**	**2.23e-10**	**6.81e-11**
119	**2.06e-07**	**7.93e-08**	**2.95e-08**	**1.06e-08**	**3.69e-09**	**1.24e-09**	**4.01e-10**	**1.25e-10**
120	**3.20e-07**	**1.26e-07**	**4.78e-08**	**1.76e-08**	**6.24e-09**	**2.14e-09**	**7.09e-10**	**2.27e-10**

TABLE B-11 Risk That ≥ j Sources Will Continue Talking Beyond 1.34 sec Out of n Sources (Row) Combined in a Single VCC, Where j = 79 to 86 and n = 116 to 120

j (column) n (row)	≥79	≥80	≥81	≥82	≥83	≥84	≥85	≥86
116	5.37e-12	1.44e-12	3.73e-13	9.25e-14	2.20e-14	5.04e-15	1.10e-15	2.32e-16
117	**1.05e-11**	2.89e-12	7.66e-13	1.95e-13	4.79e-14	1.13e-14	2.55e-15	5.52e-16
118	**2.00e-11**	5.67e-12	1.55e-12	4.05e-13	1.02e-13	2.48e-14	5.76e-15	1.29e-15
119	**3.77e-11**	**1.10e-11**	3.06e-12	8.25e-13	2.14e-13	5.33e-14	1.28e-14	2.93e-15
120	**6.99e-11**	**2.08e-11**	5.97e-12	1.65e-12	4.39e-13	1.12e-13	2.77e-14	6.55e-15

Appendix C

M/M/1 Queuing System

This model assumes a random Poisson arrival pattern and a random (exponential) service time distribution. The arrival rate does not depend upon the number of customers in the system, and the probability of an arrival in a time interval of length $h > 0$ is given by the following where λ is the average interarrival gap:

$$e^{-\lambda h}(\lambda h) = \lambda h\left(1 - \lambda h + \frac{(\lambda h)^2}{2!} - \dots\dots\dots\dots\dots\right)$$

$$= \lambda h - (\lambda h)^2 + \frac{(\lambda h)^3}{2!} - \dots\dots + (-1)^{n+1}\frac{(\lambda h)^n}{(n-1)!} + \dots\dots\dots\dots$$

$$= (\lambda h) + 0(h)$$

Thus,

$$\lambda_n = \lambda \quad n = 0, 1, 2 \dots$$

By hypothesis, the service time distribution $W(t)$ is given by the following where u = the average service duration, t = a short time interval, and s = the time of service:

$$W_s(t) = P(s \le t) = 1 - e^{-ut}, \qquad t \ge 0$$

Hence, if a customer is receiving service, the probability of a service completion (death) in a short time interval h is given by the following:

$$1 - e^{-uh} = 1 - \left(1 - uh + \frac{(uh)^2}{2!} - \dots\dots\dots\dots\dots\dots\right) = uh + o(h)$$

Here we have the memoryless property of the exponential distribution in neglecting the service already completed.

Thus,

$$\mu_n = \mu \quad n = 1, 2, 3 \dots\dots\dots\dots\dots$$

The state transition diagram for the M/M/1 queuing system is as shown here:

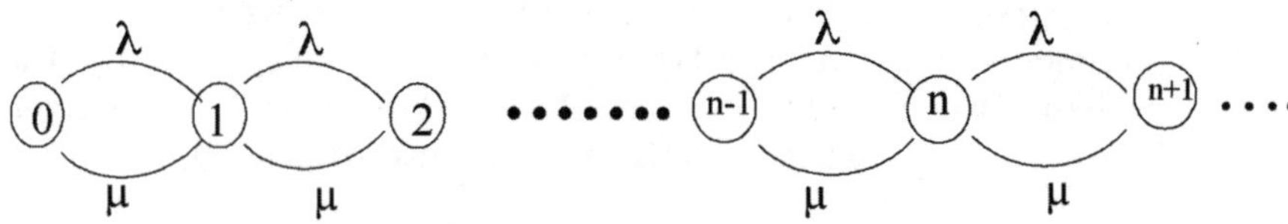

Appendix

D

Markov's Inequality Algorithm[1]

The Markov inequality method for the calculation of end-to-end delay uses the following equation:

$$cdv(\alpha) \leq -\frac{2\log(\alpha)}{\lambda} + \frac{2}{\lambda}\sum_{i=1}^{n} r_i \log\left(\frac{\lambda_i}{\lambda_i - \frac{\lambda}{2}}\right)$$

where

$$N = \text{number of switches in the path}$$

$$\lambda_i = \frac{\text{mean } CTD_i}{\text{variance } CTD_i}$$

$$r_i = \frac{(\text{mean } CTD_i)^2}{\text{variance } CTD_i}$$

$$\lambda = \min\lambda_i, \text{ over all switches } i \text{ on the path}$$

$$cdv(\alpha) = \alpha \text{ quantile of the end-to-end delay } D,$$

$$i.e\ P\{D > cdv(\alpha)\} = \alpha$$

[1] Andrei A. Markov was a graduate of Saint Petersburg University (1878), where he began as professor in 1886. Markov is particularly remembered for his study of Markov chains, sequences of random variables in which the future variable is determined by the present variable but is independent of the way in which the present state arose from its predecessors.

Appendix

E

Market Research into Voice over Alternate Networks

This appendix contains extracts from two reports commissioned by the Yankee Group in 1997 and 2001:

- In 1997, the Yankee Group Report [76] evaluated the appeal of *Voice Services over Alternate Networks* (VSANs)—specifically the public internet, frame relay, and ATM. The evaluation is based on the Yankee Communications Survey, a comprehensive research tool that surveys 300 of the Fortune 1000 companies.
- In 2001, the Yankee Group conducted the Global Networks Strategies Survey [77], an end-user telephone survey targeting large multinational corporations.

1997 Report into Voice Services

Looking first at the 1997 Yankee Group Report [76], four main drivers are identified as creators of a market opportunity for VSANs:

- Widespread deployment of data networks, both at the enterprise level and at the backbone
- Use of the Internet
- Intranet and Extranet as the leading platforms for customer, intra-, and inter-office communications
- The high cost of voice calling, especially international calling

In 1997, 7 percent of the respondents in the Yankee Communications Survey indicated that they already had some voice traffic over alternate networks. An additional 10 percent were exploring the use of VSANS on a trial basis; 55 percent were considering moving some of their voice to alternate networks; and 27 percent indicated that they were not likely to deploy voice services over alternate networks.

Potential Profitability from the Voice Services Market

The findings of the Yankee Report [76] showed the potential profitability from the voice services market. The most important market for new services is seen to be international calling. At an estimated $500 billion, the international voice services market offers the biggest opportunity for VSAN providers on a *Wide Area Network* (WAN) basis.

In many countries, the rates applied to voice services are not aligned with their costs. Monopoly markets, in general, choose to price voice services at a much higher level than competitive markets. This is especially true for international voice services, where high profits are used to subsidize local calling.

- "There is no doubt that the carriers most profitable segment, International Direct Distance Dialing (IDDD) is under threat. There is no doubt that they will—and are—losing traffic to VSANs. The billion-dollar question is how much traffic will the carriers lose? And is that loss going to be significant enough to fundamentally rethink delivery of International Direct Distance Dialing (IDDD) traffic over the Public Switched Telephone Network and circumvent the accounting rate regime."
- "We believe VSANs will significantly affect and may fundamentally alter the IDDD model over PSTN only model."

Concerns About Quality versus Cost Savings

The report [76] indicated that voice quality is still a major concern. When voice applications are considered separately, customer calling is generally regarded as one of the most mission-critical applications in an organization.

Over 80 percent of the respondents suggested that the primary reason to deploy VSANs is to lower service costs. Not a single respondent to the Yankee survey would consider putting the organization's customer calling applications on VSANs unless the savings were at least 40 percent.

Close to 30 percent of respondents expected a saving of 50 and 20 percent wanted savings as high as 80 percent if they were to deploy customer calling over alternate networks.

The Choice of Technology for the Transmission of Voice

When comparing toll quality *Public Switched Telephone Network* (PSTN) voice services with VSANs, ATM comes the closest, followed by frame relay and then IP. However, the extent of ATM deployment currently is the lowest, making voice over ATM even less of an option at the enterprise level. ATM deployment at the enterprise level is expected to grow to a point where 41 percent of large businesses will deploy some voice services over ATM.

Competition from IP

From an IP perspective, private IP networks do much better than the public internet. Over the public internet, the problem of highly variable internet delay, which significantly disturbs the natural cadence of conversation, directly affects voice quality.

A private enterprise IP network, on the other hand, ensures high availability and enables priority schemes, via a router or a *frame relay access device* (FRAD). Being a managed network, the enterprise network makes the voice quality significantly better than the public Internet.

- "A total of 75% of all users felt that the Internet's voice quality is too poor for them to move any voice traffic over the Internet and 62.5% indicated that voice over the Internet is not secure enough for their needs."
- "70% of users felt that they do not see significant savings by moving voice traffic to the Internet."

Potential for FAX and Voice Mail

Fax appears to do much better than voice as a candidate for VSANs.

- "We believe the days of using PSTN voice circuits for fax services are coming to an end."

Unlike conventional voice, fax delivery is not always expected to be immediate and low-end fax often takes up circuits that could be used for more mission-critical voice calling. Much like fax, voice mail services are seen by respondents as an application that can be moved from the PSTN to alternate networks—that is, voice mail is unidirectional and does not require full-duplex capability.

- "We believe that fax will be the first 'voice like' service to migrate to alternate networks."

In the Yankee survey, 44 percent of respondents indicated that they will put some fax traffic over the Internet, 22 percent over ATM, 30 percent over frame relay, and 17 percent over intranets.

The Technology Choice—ATM versus Frame Relay

There is a direct correlation between respondents intent on deploying voice over ATM and their expectation of growth of ATM for their data services. Respondents do not rank frame relay as highly for voice services. Instead they rank ATM higher. The Yankee Report quoted:

- "We feel that user expectation of higher deployment of ATM may in fact, result in their discounting frame relay as the service for voice." "Alternatively it could be that they plan to use ATM on high traffic routes and frame relay on low traffic routes."

2001 Report into Global Networks Strategies

While the idea of convergence of voice and data remained in the minds of end users, results of a more recent Global Network Strategies Survey carried out in October 2001 by the Yankee Group indicated that true convergence had not yet arrived.

Figure E-1 [77], compiled in 2001, shows that 66 percent of voice traffic continues to use the PSTN with 53 percent of total respondents using the PSTN for all of their voice traffic. This percentage decreased when the survey participants looked forward to 2003, with PSTN traffic down to 49 percent, and the percentage of respondents using PSTN for all of their voice traffic, also reducing to 22 percent.

An increase is anticipated in *Voice-over-IP* (VoIP) traffic from 5 percent in 2000 to 18 percent in 2003. However, the report shows that some concerns remained among end users about *Quality of Service* (QoS) and other performance issues.

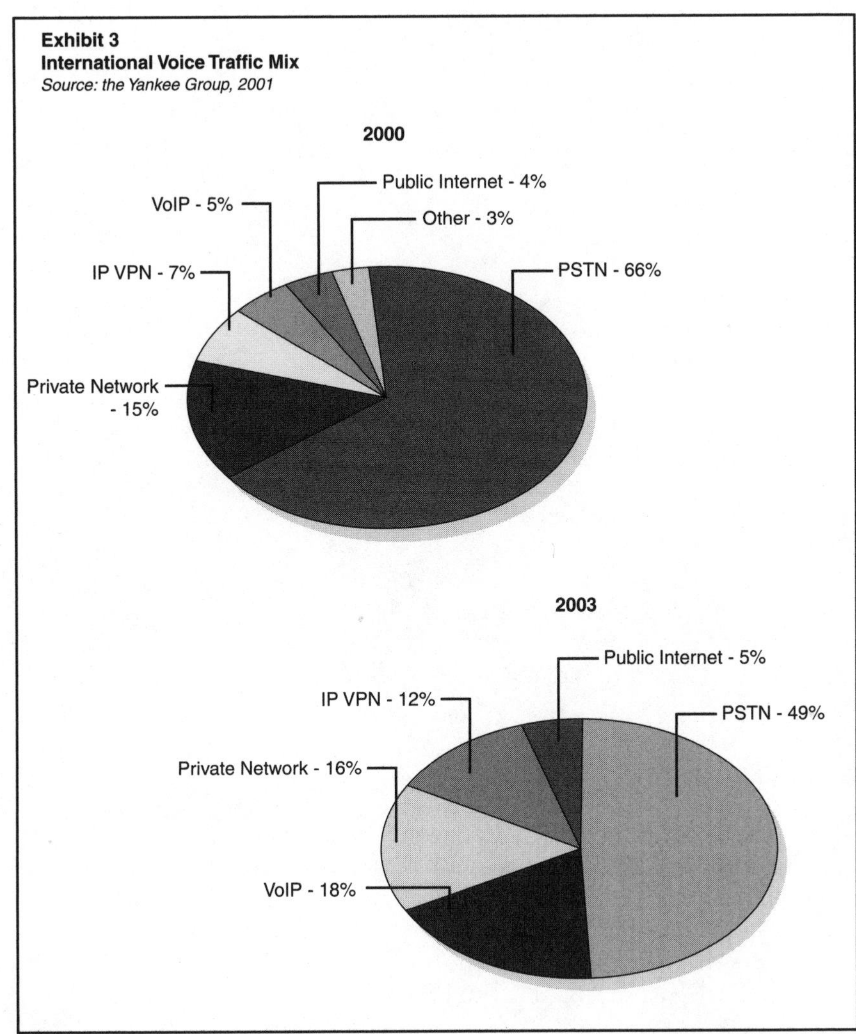

Figure E-1 International Voice Traffic Mix in 2000 and 2003

Appendix F

Digital Private Network Signaling Subsystem (DPNSS)

The principle of DPNSS is to establish a dedicated signaling channel (time slot 16) and use common channel signaling to support the 30-voice channels coexisting in the 2-Mbps link. The protocol has been in force from the mid- 1980's and has developed to support a whole host of supplementary services as can be seen from the following list.

DPNSS Supplementary Services

Circuit switched data call	Swap
Call back when free	Executive intrusion
Diversion immediate	Diversion on busy
Diversion on no reply	Hold
Three-party call	Call offer
Nonspecified information	Service independent strings
Call waiting	Bearer service selection
Route optimization	Extension status
Controlled diversion	Redirection
Series call	Three-party takeover
Night service	Centralized operator
Traffic channel maintenance	Remote alarm reporting

Add on conference
Call back when next used
Priority breakdown
Loop avoidance
Text message
Network address extension
Call distribution
Wait on busy
Traveling Class of Service
Time synchronization
Do not disturb
Call back messaging
Forced release
Charge reporting
Call park
Route capacity control
Call pick up
Number presentation restriction

The DPNSS standard is documented in a BT Network Requirement document [BTNR 188], and is available from

BT Network Standards Secretariat
BT Angel Centre
403 St John Street
London EC1V 4PL

Appendix G

QSIG

The other major inter-PBX digital signaling protocol is QSIG, which also uses common channel signaling to support the 30-voice channels coexistent in the 2-Mbps link. QSIG is rapidly gaining wide international acceptance and has been developed by *European Computer Manufacturers Association* (ECMA) specifically as a signaling system modeled on the ITU-T's Q.93x series of recommendations for basic services and generic functions, and the Q.95x services for supplementary services.

QSIG was designed to supersede proprietary private signaling systems such as DPNSS. The standards for QSIG developed by ECMA have been adopted internationally by ISO; this international private network signaling system is called *Private Signalling System 1* (PSS1).

QSIG Standards

The QSIG standards are documented as follows:

- Basic call: ECMA-106, ECMA-142, ECMA-143
- Generic support for SSs and ANFs: ECMA-156, ECMA-161, ECMA-165
- Standards for SSs and ANFs
- Identification: ECMA-148, ECMA-157
- Name identification: ECMA-163, ECMA-164
- Call diversion: ECMA-173, ECMA-174

- Path replacement: ECMA-175, ECMA-176
- Call transfer: ECMA-177, ECMA-178
- Call completion: ECMA-185, ECMA-186
- Call offer: ECMA-191, ECMA-192
- Do not disturb and override: ECMA-193, ECMA-194
- Call intrusion: ECMA-202, ECMA-203
- Advice of charge: ECMA-211, ECMA-212
- Recall: ECMA-213, ECMA-214
- Call interception: ECMA-220, ECMA-221
- Transit counter: ECMA-224, ECMA-225
- Message waiting indication: ECMA-241, ECMA-242
- Common information: ECMA-250, ECMA-251
- Call priority interruption and protection: ECMA-263, ECMA-264

ECMA has a useful web site that can be reached via the URL: http://www.ecma.ch/

A CD-ROM containing a complete collection of the ECMA Standards and Technical Reports available as files, plus the description of all ECMA Standards and Technical Reports, can be ordered by sending e-mail to documents@ecma.ch.

Further general information on QSIG can be obtained from the web pages of the IPNS forum at http://www.qsig.ie/

Appendix H

Formulas for Tables

Tables 4-1, 4-2, 5-3, 5-4, 9-1, and 9-2 are intended to be used interactively. The formulas used to build these tables in Microsoft Excel are given here:

Formula for Cell Rate Calculator for Structured Circuit Emulation Table 4-1 on page 41.

1A 2 3 4	B Number of Channels (1-31)	C Fill Level (K) (Bytes)	D Cell Rate (Cells/sec)	E Bandwidth (Kbps)
5	31	45	=ROUNDUP((((((8000*B5)/C5)* (1+1/128)+1))),0)	=D5*424/1000

Columns B and C are for variable input.

Formula for Cell Rate Calculator for T1 and E1 CAS Service, Table 4-2 on page 41.

1A 2 3 4	B Connection Type	C Number of Channels N	D Fill Rate (K) (Bytes)	E N Even	F Bandwidth (Kbps)
5	T1	24	47	=ROUNDUP(((8000*((49*C5)/(48*D5))*(1+1/128))+1),0)	=E5*424/1000
6					
7	E1	30	47	=ROUNDUP(((8000*((33*C7)/(32*D7))*(1+1/128))+1),0)	=E7*424/1000

	G N Odd	H Bandwidth (bps)
5	=ROUNDUP(((8000*(((49*C5)+1)/(48*D5))*(1+1/128))+1),0)	=G5*424/1000
6		
7	=ROUNDUP(((8000*(((33*C7)+1)/(32*D7))*(1+1/128))+1),0)	=G7*424/1000

Columns C and D are for variable input.

Formula for Bandwidth Requirements for AAL2, Table 5-3 on page 56.

A1	B	C	D	E	F	G
2	Coder	Output	Coder			Number
3		Rate of	Unit			of Channels
4		Coder				Contributing
5			(Milli-			to AAL2
6		(Bits/Sec)	secs)			Interface
7	1	2	3			4
8	Clear channel	64,000	5	=(C8/1000)*D8	=(E8/8)+3	10
9						
10	G.711 u-law	64,000	5	=(C10/1000)*D10	=(E10/8)+3	10
11	G.711 A-law	64,000	5	=(C11/1000)*D11	=(E11/8)+3	30
12						
13	G.723.1 5.3 kbps	5,300	30	=(C13/1000)*D13	=(E13/8)+3	10
14	G.723.1 6.3 kbps	6,300	30	=(C14/1000)*D14	=(E14/8)+3	30
15						
16	G.726 16 kbps	16,000	5	=(C16/1000)*D16	=(E16/8)+3	10
17	G.726 24 kbps	24,000	5	=(C17/1000)*D17	=(E17/8)+3	10
18	G.726 32 kbps	32,000	5	=(C18/1000)*D18	=(E18/8)+3	10
19	G.726 48 kbps	48,000	5	=(C19/1000)*D19	=(E19/8)+3	30
20						
21	G.729A 8 kbps	8,000	10	=(C21/1000)*D21	=(E21/8)+3	10
22						

Columns G and I are for variable input.

Formula for Bandwidth Requirements for AAL2, Table 5-3 on page 56. ***(continued)***

	H	I Choose Max.Timer setting (Millisecs) 5	J Recomm. Timer Setting for Full Cell Fill 6	K
8	=IF(G8=1,1,G8-1)	5	=(O8-1)*P8	=47-F8
9				
10	=IF(G10=1,1,G10-1)	5	=(O10-1)*P10	=47-F10
11	=IF(G11=1,1,G11-1)	5	=(O11-1)*P11	=47-F11
12				
13	=IF(G13=1,1,G13-1)	5	=(O13-1)*P13	=47-F13
14	=IF(G14=1,1,G14-1)	5	=(O14-1)*P14	=47-F14
15				
16	=IF(G16=1,1,G16-1)	5	=(O16-1)*P16	=47-F16
17	=IF(G17=1,1,G17-1)	5	=(O17-1)*P17	=47-F17
18	=IF(G18=1,1,G18-1)	5	=(O18-1)*P18	=47-F18
19	=IF(G19=1,1,G19-1)	5	=(O19-1)*P19	=47-F19
20				
21	=IF(G21=1,1,G21-1)	5	=(O21-1)*P21	=47-F21
22				

Formula for Bandwidth Requirements for AAL2, Table 5-3 on page 56. *(continued)*

	L Bandwidth Required for all Channels (Kbps) 7	M Average Bandwidth per single Channel (Kbps) 8	N % Efficiency based on contributing channels & Full Cell Fill
8	=IF((I8<P8),(((1000/P8)/U8)*53*8/1000),((((F8/47)*(1000/P8))*53*8/1000)))	=L8/G8	=(IF(I8<J8,(F8*U8)/47,1))
9			
10	=IF((I10<P10),(((1000/P10)/U10)*53*8/1000),((((F10/47)*(1000/P10))*53*8/1000)))	=L10/G10	=(IF(I10<J10,(F10*U10)/47,1))
11	=IF((I11<P11),(((1000/P11)/U11)*53*8/1000),((((F11/47)*(1000/P11))*53*8/1000)))	=L11/G11	=(IF(I11<J11,(F11*U11)/47,1))
12			
13	=IF((I13<P13),(((1000/P13)/U13)*53*8/1000),((((F13/47)*(1000/P13))*53*8/1000)))	=L13/G13	=(IF(I13<J13,(F13*U13)/47,1))
14	=IF((I14<P14),(((1000/P14)/U14)*53*8/1000),((((F14/47)*(1000/P14))*53*8/1000)))	=L14/G14	=(IF(I14<J14,(F14*U14)/47,1))
15			
16	=IF((I16<P16),(((1000/P16)/U16)*53*8/1000),((((F16/47)*(1000/P16))*53*8/1000)))	=L16/G16	=(IF(I16<J16,(F16*U16)/47,1))
17	=IF((I17<P17),(((1000/P17)/U17)*53*8/1000),((((F17/47)*(1000/P17))*53*8/1000)))	=L17/G17	=(IF(I17<J17,(F17*U17)/47,1))
18	=IF((I18<P18),(((1000/P18)/U18)*53*8/1000),((((F18/47)*(1000/P18))*53*8/1000)))	=L18/G18	=(IF(I18<J18,(F18*U18)/47,1))
19	=IF((I19<P19),(((1000/P19)/U19)*53*8/1000),((((F19/47)*(1000/P19))*53*8/1000)))	=L19/G19	=(IF(I19<J19,(F19*U19)/47,1))
20			
21	=IF((I21<P21),(((1000/P21)/U21)*53*8/1000),((((F21/47)*(1000/P21))*53*8/1000)))	=L21/G21	=(IF(I21<J21,(F21*U21)/47,1))
22			

Formula for Bandwidth Requirements for AAL2, Table 5-3 on page 56. ***(continued)***

	O Number of channels required for Full Cell Fill	P Average sample gap	Q Filler in a cell ? based on bandwidth	R Average sample rate/rec.timer value millisecs
8	=ROUNDUP((47/F8),0)	=IF(V8>R8,R8,V8)	=IF((S8<O8),47-(S8*F8),0)	=D8/G8
9				
10	=ROUNDUP((47/F10),0)	=IF(V10>R10,R10,V10)	=IF((S10<O10),47-(S10*F10),0)	=D10/G10
11	=ROUNDUP((47/F11),0)	=IF(V11>R11,R11,V11)	=IF((S11<O11),47-(S11*F11),0)	=D11/G11
12				
13	=ROUNDUP((47/F13),0)	=IF(V13>R13,R13,V13)	=IF((S13<O13),47-(S13*F13),0)	=D13/G13
14	=ROUNDUP((47/F14),0)	=IF(V14>R14,R14,V14)	=IF((S14<O14),47-(S14*F14),0)	=D14/G14
15				
16	=ROUNDUP((47/F16),0)	=IF(V16>R16,R16,V16)	=IF((S16<O16),47-(S16*F16),0)	=D16/G16
17	=ROUNDUP((47/F17),0)	=IF(V17>R17,R17,V17)	=IF((S17<O17),47-(S17*F17),0)	=D17/G17
18	=ROUNDUP((47/F18),0)	=IF(V18>R18,R18,V18)	=IF((S18<O18),47-(S18*F18),0)	=D18/G18
19	=ROUNDUP((47/F19),0)	=IF(V19>R19,R19,V19)	=IF((S19<O19),47-(S19*F19),0)	=D19/G19
20				
21	=ROUNDUP((47/F21),0)	=IF(V21>R21,R21,V21)	=IF((S21<O21),47-(S21*F21),0)	=D21/G21
22				

Formula for Bandwidth Requirements for AAL2, Table 5-3 on page 56. *(continued)*

	S Final possible Number of channels in one cell	T Final samples in 1 cell within timer	U Number of Samples is never Zero	V Inter sample gap subject to timer constraint
8	=IF((W8+1)<O8,W8+1,O8)	=IF((I8<R8),ROUNDDOWN(I8/R8,0),S8)	=IF(T8=0,1,T8)	=D8/S8
9				
10	=IF((W10+1)<O10,W10+1,O10)	=IF((I10<R10),ROUNDDOWN(I10/R10,0),S10)	=IF(T10=0,1,T10)	=D10/S10
11	=IF((W11+1)<O11,W11+1,O11)	=IF((I11<R11),ROUNDDOWN(I11/R11,0),S11)	=IF(T11=0,1,T11)	=D11/S11
12				
13	=IF((W13+1)<O13,W13+1,O13)	=IF((I13<R13),ROUNDDOWN(I13/R13,0),S13)	=IF(T13=0,1,T13)	=D13/S13
14	=IF((W14+1)<O14,W14+1,O14)	=IF((I14<R14),ROUNDDOWN(I14/R14,0),S14)	=IF(T14=0,1,T14)	=D14/S14
15				
16	=IF((W16+1)<O16,W16+1,O16)	=IF((I16<R16),ROUNDDOWN(I16/R16,0),S16)	=IF(T16=0,1,T16)	=D16/S16
17	=IF((W17+1)<O17,W17+1,O17)	=IF((I17<R17),ROUNDDOWN(I17/R17,0),S17)	=IF(T17=0,1,T17)	=D17/S17
18	=IF((W18+1)<O18,W18+1,O18)	=IF((I18<R18),ROUNDDOWN(I18/R18,0),S18)	=IF(T18=0,1,T18)	=D18/S18
19	=IF((W19+1)<O19,W19+1,O19)	=IF((I19<R19),ROUNDDOWN(I19/R19,0),S19)	=IF(T19=0,1,T19)	=D19/S19
20				
21	=IF((W21+1)<O21,W21+1,O21)	=IF((I21<R21),ROUNDDOWN(I21/R21,0),S21)	=IF(T21=0,1,T21)	=D21/S21
22				

Formula for Bandwidth Requirements for AAL5, Table 5-4 on page 58.

A1 2 3 4 5	B Coder Type	C Output Rate of Coder (bps)	D Coder Unit (millis)	E Packing Factor Range	F Packing Factor Selected
6	Clear channel 64 kbps	64,000	5	1 to 4	4
7					
8	G.711 u-law 64 kbps	64,000	5	1 to 4	4
9	G.711 A-law 64 kbps	64,000	5	1 to 4	2
10					
11	G.723.1 5.3 kbps	5,300	30	1 to 6	4
12	G.723.1 6.3 kbps	6,300	30	1 to 7	1
13					
14	G.726 16 kbps	16,000	5	1 to 18	4
15	G.726 24 kbps	24,000	5	1 to 15	1
16	G.726 32 kbps	32,000	5	1 to 9	3
17	G.726 40 kbps	40,000	5	1 to 7	7
18					
19	G.729A 8 kbps	8,000	10	1 to 13	13
20					

Column F is for variable input.

Formula for Bandwidth Requirements for AAL5, Table 5-4 on page 58.

	G Number of ATM cells Required incl. Overheads	H Bandwidth Required incl. Overhead (Kbps)
6	=ROUNDUP((((C6/1000)*D6*F6)+80)/384,0)	=G6*424/(D6*F6)
7		
8	=ROUNDUP((((C8/1000)*D8*F8)+80)/384,0)	=G8*424/(D8*F8)
9	=ROUNDUP((((C9/1000)*D9*F9)+80)/384,0)	=G9*424/(D9*F9)
10		
11	=ROUNDUP((((C11/1000)*D11*F11)+80)/384,0)	=G11*424/(D11*F11)
12	=ROUNDUP((((C12/1000)*D12*F12)+80)/384,0)	=G12*424/(D12*F12)
13		
14	=ROUNDUP((((C14/1000)*D14*F14)+80)/384,0)	=G14*424/(D14*F14)
15	=ROUNDUP((((C15/1000)*D15*F15)+80)/384,0)	=G15*424/(D15*F15)
16	=ROUNDUP((((C16/1000)*D16*F16)+80)/384,0)	=G16*424/(D16*F16)
17	=ROUNDUP((((C17/1000)*D17*F17)+80)/384,0)	=G17*424/(D17*F17)
18		
19	=ROUNDUP((((C19/1000)*D19*F19)+80)/384,0)	=G19*424/(D19*F19)

Formula for Table 9-1: An average talkspurt size (1.34 s and 2 s) in cells for 10 coder types using AAL5, on page 116.

Column No.	A Coder Type	B Output rate of coder (bps)	C Coder Unit	D Packing Factor Range	E Packing Factor Selected	F Number of ATM cells incl. Overheads
6	Clear channel 64 kbps	64,000	5	1 to 4	**2**	=ROUNDUP((((B6/1000)*C6*E6)+80)/384,0)
7						
8	G.711 u-law 64 kbps	64,000	5	1 to 4	**2**	=ROUNDUP((((B8/1000)*C8*E8)+80)/384,0)
9	G.711 A-law 64 kbps	64,000	5	1 to 4	**1**	=ROUNDUP((((B9/1000)*C9*E9)+80)/384,0)
10						
11	G.723.1 5.3 kbps	5,300	30	1 to 6	**3**	=ROUNDUP((((B11/1000)*C11*E11)+80)/384,0)
12	G.723.1 6.3 kbps	6,300	30	1 to 7	**7**	=ROUNDUP((((B12/1000)*C12*E12)+80)/384,0)
13						
14	G.726 16 kbps	16,000	5	1 to 18	**4**	=ROUNDUP((((B14/1000)*C14*E14)+80)/384,0)
15	G.726 24 kbps	24,000	5	1 to 15	**2**	=ROUNDUP((((B15/1000)*C15*E15)+80)/384,0)
16	G.726 32 kbps	32,000	5	1 to 9	**3**	=ROUNDUP((((B16/1000)*C16*E16)+80)/384,0)
17	G.726 40 kbps	40,000	5	1 to 7	**1**	=ROUNDUP((((B17/1000)*C17*E17)+80)/384,0)
18						
19	G.729A 8 kbps	8,000	10	1 to 13	**10**	=ROUNDUP((((B19/1000)*C19*E19)+80)/384,0)

Column E is for variable input.

Formula for Table 9-1: An average talkspurt size (1.34 s and 2 s) in cells for 10 coder types using AAL5, on page 116.

	G Bandwidth Required inc. Overhead (Kbps)	H Number of Cells in a talkspurt of 1.34 sec	I Number of Cells in a talkspurt of 2.0 sec
6	=F6*424/(C6*E6)	=(1340/(E6*C6)*F6)+1	=(2000/(E6*C6)*F6)+1
7			
8	=F8*424/(C8*E8)	=(1340/(E8*C8)*F8)+1	=(2000/(E8*C8)*F8)+1
9	=F9*424/(C9*E9)	=(1340/(E9*C9)*F9)+1	=(2000/(E9*C9)*F9)+1
10			
11	=F11*424/(C11*E11)	=(1340/(E11*C11)*F11)+1	=(2000/(E11*C11)*F11)+1
12	=F12*424/(C12*E12)	=(1340/(E12*C12)*F12)+1	=(2000/(E12*C12)*F12)+1
13			
14	=F14*424/(C14*E14)	=(1340/(E14*C14)*F14)+1	=(2000/(E14*C14)*F14)+1
15	=F15*424/(C15*E15)	=(1340/(E15*C15)*F15)+1	=(2000/(E15*C15)*F15)+1
16	=F16*424/(C16*E16)	=(1340/(E16*C16)*F16)+1	=(2000/(E16*C16)*F16)+1
17	=F17*424/(C17*E17)	=(1340/(E17*C17)*F17)+1	=(2000/(E17*C17)*F17)+1
18			
19	=F19*424/(C19*E19)	=(1340/(E19*C19)*F19)+1	=(2000/(E19*C19)*F19)+1

Formula for Table 9-2: An average talkspurt size (1.34 s and 2 s) in cells for 10 coder types using AAL2, on page 117.

Column No.	A Coder	B Output Rate of Coder (bps)	C Coder Unit (Milli-secs)	D Number of bits in one sample	E Number of bytes in one sample	F Number of Channels Contributing to AAL2 Interface	G
8	Clear channel	64,000	5	=(B8/1000)*C8	=(D8/8)+3	2	=IF(F8=1,1,F8-1)
9							
10	G.711 u-law	64,000	5	=(B10/1000)*C10	=(D10/8)+3	2	=IF(F10=1,1,F10-1)
11	G.711 A-law	64,000	5	=(B11/1000)*C11	=(D11/8)+3	2	=IF(F11=1,1,F11-1)
12							
13	G.723.1 5.3 kbps	5,300	30	=(B13/1000)*C13	=(D13/8)+3	2	=IF(F13=1,1,F13-1)
14	G.723.1 6.3 kbps	6,300	30	=(B14/1000)*C14	=(D14/8)+3	2	=IF(F14=1,1,F14-1)
15							
16	G.726 16 kbps	16,000	5	=(B16/1000)*C16	=(D16/8)+3	2	=IF(F16=1,1,F16-1)
17	G.726 24 kbps	24,000	5	=(B17/1000)*C17	=(D17/8)+3	2	=IF(F17=1,1,F17-1)
18	G.726 32 kbps	32,000	5	=(B18/1000)*C18	=(D18/8)+3	2	=IF(F18=1,1,F18-1)
19	G.726 40 kbps	40,000	5	=(B19/1000)*C19	=(D19/8)+3	2	=IF(F19=1,1,F19-1)
20							
21							
22	G.729A 8 kbps	8,000	10	=(B22/1000)*C22	=(D22/8)+3	2	=IF(F22=1,1,F22-1)

Column F is for variable input.

Formula for Table 9-2: An average talkspurt size (1.34 s and 2 s) in cells for 10 coder types using AAL2, on page 117.

	H Choose Max.Timer setting (Millisecs)	I Recommended Timer Setting for Full Cell Fill	J
8	3	=ROUNDUP((P8-1)*Q8,2)	=47-E8
9			
10	3	=ROUNDUP((P10-1)*Q10,2)	=47-E10
11	3	=ROUNDUP((P11-1)*Q11,2)	=47-E11
12			
13	3	=ROUNDUP((P13-1)*Q13,2)	=47-E13
14	3	=ROUNDUP((P14-1)*Q14,2)	=47-E14
15			
16	3	=ROUNDUP((P16-1)*Q16,2)	=47-E16
17	3	=ROUNDUP((P17-1)*Q17,2)	=47-E17
18	3	=ROUNDUP((P18-1)*Q18,2)	=47-E18
19	3	=ROUNDUP((P19-1)*Q19,2)	=47-E19
20			
21			
22	3	=ROUNDUP((P22-1)*Q22,2)	=47-E22

Column H is for variable input.

Formula for Table 9-2: An average talkspurt size (1.34 s and 2 s) in cells for 10 coder types using AAL2, on page 117.

	K Bandwidth Required for all Channels (Kbps)	L Average Bandwidth per single Channel (Kbps)
8	=IF((H8<Q8),(((1000/Q8)/V8)*53*8/1000),((((E8/47)*(1000/Q8))*53*8/1000)))	=K8/F8
9		
10	=IF((H10<Q10),(((1000/Q10)/V10)*53*8/1000),((((E10/47)*(1000/Q10))*53*8/1000)))	=K10/F10
11	=IF((H11<Q11),(((1000/Q11)/V11)*53*8/1000),((((E11/47)*(1000/Q11))*53*8/1000)))	=K11/F11
12		
13	=IF((H13<Q13),(((1000/Q13)/V13)*53*8/1000),((((E13/47)*(1000/Q13))*53*8/1000)))	=K13/F13
14	=IF((H14<Q14),(((1000/Q14)/V14)*53*8/1000),((((E14/47)*(1000/Q14))*53*8/1000)))	=K14/F14
15		
16	=IF((H16<Q16),(((1000/Q16)/V16)*53*8/1000),((((E16/47)*(1000/Q16))*53*8/1000)))	=K16/F16
17	=IF((H17<Q17),(((1000/Q17)/V17)*53*8/1000),((((E17/47)*(1000/Q17))*53*8/1000)))	=K17/F17
18	=IF((H18<Q18),(((1000/Q18)/V18)*53*8/1000),((((E18/47)*(1000/Q18))*53*8/1000)))	=K18/F18
19	=IF((H19<Q19),(((1000/Q19)/V19)*53*8/1000),((((E19/47)*(1000/Q19))*53*8/1000)))	=K19/F19
20		
21		
22	=IF((H22<Q22),(((1000/Q22)/V22)*53*8/1000),((((E22/47)*(1000/Q22))*53*8/1000)))	=K22/F22

Formula for Table 9-2: An average talkspurt size (1.34 s and 2 s) in cells for 10 coder types using AAL2, on page 117.

	M % Efficiency based on Contributing Channels & Full Cell Fill	N Number of cells in a talkspurt of 1.34 secs	O Number of cells in a talkspurt of 2 secs
8	=(IF(H8<I8,(E8*V8)/47,1))	=ROUNDUP(((1340/C8*F8*(E8+3))/48),0)+1	=ROUNDUP(((2000/C8*F8*(E8+3))/48),0)+1
9			
10	=(IF(H10<I10,(E10*V10)/47,1))	=ROUNDUP(((1340/C10*F10*(E10+3))/48),0)+1	=ROUNDUP(((2000/C10*F10*(E10+3))/48),0)+1
11	=(IF(H11<I11,(E11*V11)/47,1))	=ROUNDUP(((1340/C11*F11*(E11+3))/48),0)+1	=ROUNDUP(((2000/C11*F11*(E11+3))/48),0)+1
12			
13	=(IF(H13<I13,(E13*V13)/47,1))	=ROUNDUP(((1340/C13*F13*(E13+3))/48),0)+1	=ROUNDUP(((2000/C13*F13*(E13+3))/48),0)+1
14	=(IF(H14<I14,(E14*V14)/47,1))	=ROUNDUP(((1340/C14*F14*(E14+3))/48),0)+1	=ROUNDUP(((2000/C14*F14*(E14+3))/48),0)+1
15			
16	=(IF(H16<I16,(E16*V16)/47,1))	=ROUNDUP(((1340/C16*F16*(E16+3))/48),0)+1	=ROUNDUP(((2000/C16*F16*(E16+3))/48),0)+1
17	=(IF(H17<I17,(E17*V17)/47,1))	=ROUNDUP(((1340/C17*F17*(E17+3))/48),0)+1	=ROUNDUP(((2000/C17*F17*(E17+3))/48),0)+1
18	=(IF(H18<I18,(E18*V18)/47,1))	=ROUNDUP(((1340/C18*F18*(E18+3))/48),0)+1	=ROUNDUP(((2000/C18*F18*(E18+3))/48),0)+1
19	=(IF(H19<I19,(E19*V19)/47,1))	=ROUNDUP(((1340/C19*F19*(E19+3))/48),0)+1	=ROUNDUP(((2000/C19*F19*(E19+3))/48),0)+1
20			
21			
22	=(IF(H22<I22,(E22*V22)/47,1))	=ROUNDUP(((1340/C22*F22*(E22+3))/48),0)+1	=ROUNDUP(((2000/C22*F22*(E22+3))/48),0)+1

Formula for Table 9-2: An average talkspurt size (1.34 s and 2 s) in cells for 10 coder types using AAL2, on page 117.

	P Number of channels required for Full Cell Fill	Q Average Sample Gap	R Filler in a Cell ? Based on Bandwidth	S Average Sample Rate/rec.timer Value Millisecs
8	=ROUNDUP((47/F8),0)	=IF(V8>R8,R8,V8)	=IF((S8<O8),47-(S8*F8),0)	=D8/G8
9				
10	=ROUNDUP((47/F10),0)	=IF(V10>R10,R10,V10)	=IF((S10<O10),47-(S10*F10),0)	=D10/G10
11	=ROUNDUP((47/F11),0)	=IF(V11>R11,R11,V11)	=IF((S11<O11),47-(S11*F11),0)	=D11/G11
12				
13	=ROUNDUP((47/F13),0)	=IF(V13>R13,R13,V13)	=IF((S13<O13),47-(S13*F13),0)	=D13/G13
14	=ROUNDUP((47/F14),0)	=IF(V14>R14,R14,V14)	=IF((S14<O14),47-(S14*F14),0)	=D14/G14
15				
16	=ROUNDUP((47/F16),0)	=IF(V16>R16,R16,V16)	=IF((S16<O16),47-(S16*F16),0)	=D16/G16
17	=ROUNDUP((47/F17),0)	=IF(V17>R17,R17,V17)	=IF((S17<O17),47-(S17*F17),0)	=D17/G17
18	=ROUNDUP((47/F18),0)	=IF(V18>R18,R18,V18)	=IF((S18<O18),47-(S18*F18),0)	=D18/G18
19	=ROUNDUP((47/F19),0)	=IF(V19>R19,R19,V19)	=IF((S19<O19),47-(S19*F19),0)	=D19/G19
20				
21				
22	=ROUNDUP((47/F22),0)	=IF(V22>R22,R22,V22)	=IF((S22<O22),47-(S22*F22),0)	=D22/G22

Formula for Table 9-2: An average talkspurt size (1.34 s and 2 s) in cells for 10 coder types using AAL2, on page 117.

	T Final Possible Number of channels in One Cell	U Final Samples in 1 Cell Within Timer	V Number of Samples Is Never Zero	W Inter Sample Gap Subject to Timer Constraint
8	=IF((W8+1)<O8,W8+1,O8)	=IF((I8<R8),ROUNDDOWN(I8/R8,0),S8)	=IF(T8=0,1,T8)	=D8/S8
9				
10	=IF((W10+1)<O10,W10+1,O10)	=IF((I10<R10),ROUNDDOWN(I10/R10,0),S10)	=IF(T10=0,1,T10)	=D10/S10
11	=IF((W11+1)<O11,W11+1,O11)	=IF((I11<R11),ROUNDDOWN(I11/R11,0),S11)	=IF(T11=0,1,T11)	=D11/S11
12				
13	=IF((W13+1)<O13,W13+1,O13)	=IF((I13<R13),ROUNDDOWN(I13/R13,0),S13)	=IF(T13=0,1,T13)	=D13/S13
14	=IF((W14+1)<O14,W14+1,O14)	=IF((I14<R14),ROUNDDOWN(I14/R14,0),S14)	=IF(T14=0,1,T14)	=D14/S14
15				
16	=IF((W16+1)<O16,W16+1,O16)	=IF((I16<R16),ROUNDDOWN(I16/R16,0),S16)	=IF(T16=0,1,T16)	=D16/S16
17	=IF((W17+1)<O17,W17+1,O17)	=IF((I17<R17),ROUNDDOWN(I17/R17,0),S17)	=IF(T17=0,1,T17)	=D17/S17
18	=IF((W18+1)<O18,W18+1,O18)	=IF((I18<R18),ROUNDDOWN(I18/R18,0),S18)	=IF(T18=0,1,T18)	=D18/S18
19	=IF((W19+1)<O19,W19+1,O19)	=IF((I19<R19),ROUNDDOWN(I19/R19,0),S19)	=IF(T19=0,1,T19)	=D19/S19
20				
21				
22	=IF((W22+1)<O22,W22+1,O22)	=IF((I22<R22),ROUNDDOWN(I22/R22,0),S22)	=IF(T22=0,1,T22)	=D22/S22

Glossary

A-law A companding technique used to compress voice signals. This is generally used everywhere except for North America and Japan.

Asynchronous Digital Subscriber Line (ADSL) A method of modulation of the signal over the existing POTS connection, providing higher bandwidth, typically 512 Kbps at the time of writing.

Asynchronous Transfer Mode (ATM) A switching/ transmission technology that employs 53-byte cells as a basic unit of transfer.

ATM Adaptation Layer (AAL) A layer of the ATM protocol stack used to convert non-ATM information to and from ATM cells.

ATM End System Address (AESA) The ATM addressing scheme that allows three NSAP addressing formats (E.164, DCC, and ICD) to be encoded based on ISO 8348.

ATM InterNetwork Interface (AINI) A dynamic routing protocol over static routes. This is generally used between networks using PNNI.

Available Bit Rate (ABR) A standard ATM service category.

Base Station (BS) A radio base station used to transport GSM voice and low-speed data services.

Base Station Concentrator (BSC) A concentrator connecting several mobile base stations.

Bearer Independent Call Control (BICC) A control protocol that can be used between MSCs when controlling media gateways.

Broadband Integrated Services Digital Network (B-ISDN) An ISDN that provides transmission channels capable of supporting rates greater than 2 Mbps. It uses ATM as the transfer mode.

Broadband ISDN User Part (B-ISUP) Provides the signaling functions necessary to set up, manage, and terminate a virtual circuit in a broadband network. This is based on its narrowband counterpart ISUP.

Cell Delay Variation Tolerance (CDVT) A traffic parameter that must be specified for each connection, but is not signaled. A measure of the maximum number of back-to-back cells that can be generated at line rate and still be considered to be compliant by the ingress policing algorithm.

Channel Associated Signaling (CAS) Periodic transmission of AB or ABCD signaling bit patterns rather than signaling messages. Even though the patterns are transmitted in a common timeslot for E1, it is called CAS rather than CCS because it is not message based. In CAS on an E1 link, 4 bits of signaling (ABCD) per timeslot are multiplexed into a fixed portion of timeslot 16 as defined in G.732. This mode is also known as PCM 30. In CAS on a T1 link, a single bit of signaling for a timeslot is written into the least significant bit of every sixth byte of that timeslot.

Common Channel Signaling (CCS) A message-oriented signaling protocol. Signaling messages for channels from one or more T1/E1 ports are communicated over a single common signaling channel. Examples of CCS protocols include Q.931, Q.SIG, Cornet, and DPNSS.

Circuit Emulation Service (CES) The ATM Forum circuit emulation service interoperability specification specifies interoperability agreements for supporting Constant Bit Rate (CBR) traffic over ATM networks. Specifically, this specification supports the emulation of existing TDM circuits over ATM networks.

Connection Admission Control (CAC) A function that evaluates whether the admission of a new connection maintains the Quality of Service (QoS) for all connections of that traffic class, rejecting the connection if congestion is anticipated.

Cyclic Redundancy Check (CRC) A method of checking the integrity of transmitted data.

Data Link Connection Identifier (DLCI) Identifies a frame relay connection.

Digital Service Level 0 (DS0) A worldwide standard speed of 64 Kbps for transmitting PCM digitized voiceband signals.

Digital Service Level 1 (DS1) The primary-rate digital data stream that runs at 1.544 Mbps in North America and at 2.048 Mbps in Europe.

Digital Subscriber Signaling System #1 (DSS1) A narrowband signaling protocol defined by ITU-T Q.931 that specifies the basic call control of UNI layer 3. ITU-T Q.921 specifies the DSS1 UNI layer 2.

Digital Subscriber Signaling System #2 (DSS2) A broadband signaling protocol defined by ITU-T Q.2931 that specifies the basic call/connection control of UNI layer 3. ITU-T Q.2932.1 specifies the DSS2 generic functional protocol of UNI layer 3.

Dual Tone Multiple Frequency (DTMF) DTMF address signaling is a method of signaling using the speech transmission path. This method employs 16 distinct signals, each composed of two voiceband frequencies. These 16-tone pairs may also be called touch-tone signals. DTMF is the usual method of address signaling for POTS. DTMF tones are generated by telephone keypads and PBXs to indicate the called number.

E1 A level 1 digital trunk operating at 2.048 Mbps that is used outside of North America and Japan. It is made up of 32 DS0 channels. Channel 0 is reserved for framing and alarm information, and Channel 16 is usually used for signaling, leaving 30 channels to carry data.

E.164 A telephony numbering plan (ITU-T E.164) that specifies ISDN numbers and is used as an ATM address to identify a network point of attachment. An E.164 number may be formatted as either a native address or encapsulated within an AESA.

Echo Cancellation or Echo Canceller (EC) Echo cancellation refers to the processing performed on a signal to remove unwanted echoes. An echo canceller is a device that performs such processing.

Echo Return Loss (ERL) The attenuation of echo due to transmission and hybrid loss in the near-end path associated with an echo canceller.

Echo Return Loss Enhancement (ERLE) The attenuation of the external echo signal as it passes through the echo canceller toward the talker, excluding nonlinear processing by the echo canceller.

Effective bandwidth Effective bandwidth, also sometimes known as virtual bandwidth, is expressed dynamically as the bandwidth and buffer space required to service an ATM source. Effective bandwidth is derived from the PCR, SCR, and MBS parameters and conforms to the target CLR and CDV of the ATM QoS class.

Egress Flowing in a direction away from the ATM switch core.

End System Identifier (ESI) An AESA component that can be a global unique identifier of an end system and may be assigned statically or configured dynamically by way of ILMI address registration mechanism with the subscriber terminal.

European Telecommunications Standards Institute (ETSI) A standards body—UMTS was initially developed within ETSI.

Gateway GPRS Support Node (GGSN) In the UTRAN, packetized data travels to the IP backbone via an SGSN and a GGSN.

General Packet Radio Service (GPRS) The starting point, in the building of the third-generation UMTS architecture, is the introduction of a modified Internet protocol, GPRS, preparing the circuit-switched GSM networks for packet switching.

Global System for Mobile Communications (GSM) Current GSM networks support voice and low-speed data services that are circuit switched—that is, the traffic is carried between users in bearer circuits that are switched under the control of signaling from the users.

Ingress Flowing in a direction toward the ATM switch core.

Integrated (used to be Interim) Local Management Interface (ILMI) A protocol used for the configuration and control of the locally significant parameters—for example, the network prefix of the local switch—for each ATM physical link.

Integrated Services Digital Network (ISDN) A network that provides or supports a range of different telecommunication services using digital connections.

Interim Interswitch Signaling Protocol (IISP) A signaling protocol that is built upon the ATM Forum UNI 3.1 specification that allows for some base-level capability at the private NNI interface.

International Code Designator (ICD) An AESA format that identifies an internationally significant organization that has assigned the address.

International Mobile Telecommunications 2000 Project (IMT 2000) IMT 2000 is an initiative of the International Telecommunications Union (ITU) to provide wireless access to the global telecommunication infrastructure, through both satellite and terrestrial systems, serving fixed and mobile users in public and private networks.

Internet Protocol (IP) A layer 3 connectionless protocol for packet transport.

ISDN User Part (ISUP) Provides the signaling functions necessary to set up, manage, and terminate circuits in a narrowband network.

Iu-CS The interface for circuit-switched traffic between an RNC and the core network.

Iu-PS The interface for packet-switched traffic between an RNC and the core network.

Mobile Switching Center (MSC) An MSC may interconnect BSCs and RNCs. An MSC may also act as a gateway between the circuit-switched domain and the PSTN.

Multiple Subscriber Number (MSN) An MSN is an address (or address prefix) that represents multiple subscribers.

National ISDN-2 (NI-2) A specification for a variant (subset) of Q.931 signaling that is used in North America. This is standardized by Bellcore.

Narrowband Integrated Services Digital Network (N-ISDN) An ISDN that provides transmission channels capable of 2 Mbps.

Network Node Interface (NNI) The interface between two ATM network nodes.

Network Service Access Point (NSAP) An OSI address format that has been adopted by the ATM Forum to represent a private network point of attachment and is also referred to as an AESA.

Permanent Virtual Connection (PVC) An end-to-end connection.

Permanent Virtual Path (PVP) A "bundle" of permanent virtual connections.

Plain Old Telephone System (POTS)

Primary Rate Interface (PRI) A T1 primary interface provides 23 voice channels. An E1 primary interface provides 30 voice channels. Both types need an additional two channels for signaling.

Private Integrated Services Signaling System #1 (PSS1) An internationally used set of standards that was developed by the ISO for QSIG signaling between digital PBXs.

Private Branch eXchange (PBX) A telephone exchange. In this book, transmitting PBXs are assumed to be providing a digital signal to the ATM switch.

Private Integrated Services Signaling System #1 (PSS1) An internationally used set of standards that was developed by the ISO for QSIG signaling between digital PBXs.

Private Network-to-Network Interface (P-NNI) A set of protocols used between private ATM switches. P-NNI includes a protocol for distributing topology information between switches and clusters of switches, which is used to compute paths through the network, and a signaling protocol used to establish point-to-point or point-to-multipoint connections across the ATM network. The signaling protocol is based on ATM Forum UNI signaling with extensions to support P-NNI functions.

Protocol Data Unit (PDU) A unit of transfer between layers of the ATM protocol stack.

Public Switched Telephone Network (PSTN)

Pulse Code Modulation (PCM) A method of coding for digitizing analog voice signals.

Q.2931 An ITU-T Recommendation that specifies the procedures for the establishing, maintaining, and clearing of network connections at the B-ISDN (ATM) UNI. This recommendation specifies the layer 3 call/connection states, messages, information elements, timers, and procedures used for the control of B-ISDN point-to-point on-demand calls on virtual channels.

Quality of Service (QoS) A particular instance of a set of generic parameters that are used to describe intrinsic traffic characteristics of an ATM connection.

Radio Access Network (RAN)

Radio Network Controller (RNC) In a UMTS network, an RNC controls the interface between the radio access and the ATM/IP transport in the core.

RAN Application Part (RANAP)

Real-time Variable Bit Rate (Rt-VBR) A standard ATM service category.

Routing Control Channel (RCC) The VCC that is used by P-NNI nodes to communicate routing information to each other. It is also known as a P-NNI link.

Segmentation and Reassembly (SAR) In the ATM protocol stack, the SAR sublayer provides service to the adaptation layer below. The job of the SAR sublayer is to segment the outgoing PDU into cells and reassemble the incoming PDU back into an AAL frame.

Serving GPRS Support Node (SGSN) In the UTRAN, packetized data travels to the IP backbone via an SGSN and a GGSN.

Signaling at the Q Reference Point (QSIG) A private ISDN peer-to-peer protocol used between PBXs. It is based on and is very similar to Q.931.

Signaling System 7 (SS7) The high-speed, digital common channel signaling network required for ISDN applications; it also provides myriad services based upon the calling party's number. This is described in the ITU-T Q.700 series of recommendations. The U.S. national version is described in ANSI T1.100 series standards.

Single Subscriber Number (SSN) An SSN is an address that represents a single subscriber.

Soft Permanent Virtual Connection (SPVC) The SPVC endpoints are selected by the network operator. The actual path of the SPVC through the network is dependent on the availability of network resources.

Switched Virtual Connection (SVC) A connection that is set up on demand via a signaling protocol. Such connections tend to be of shorter duration than PVCs. SVCs are not automatically reestablished after a system restarts.

Synchronous Digital Hierarchy (SDH) A digital transport network using an STM as a basic transport structure to generate a range of higher transport structures.

T1 An ISDN interface operating at 1.544 Mbps that is used inside North America and Canada. It is made up of 24 DS0 channels. One extra bit is available for framing. CAS is embedded through robbed signaling. CCS uses channel 24, leaving 23 bearer channels.

UMTS Terrestrial Radio Access Network (UTRAN)

Universal Mobile Telecommunications System (UMTS) A third-generation network allowing wireless access to the global telecommunication infrastructure, through both satellite and terrestrial systems, serving fixed and mobile users in public and private networks.

Unspecified Bit Rate (UBR) A standard ATM service category.

User-Network Interface (UNI) The shared boundary between the network and the user terminal equipment across which they communicate.

Virtual Channel Connection (VCC) A communication channel that provides for the sequential transport of ATM cells.

Virtual Channel Identifier (VCI) A 16-bit value used to identify an ATM connection and carried in the cell header. The value has local significance only and as such identifies the VCC active on the local port that forms part of an end-to-end VCC.

Voice over IP (VoIP)

Virtual Path (VP) A logical association or bundle of VCCs.

Virtual Path Connection (VPC) An end-to-end ATM connection that is capable of carrying one or more VCCs.

Virtual Path Identifier (VPI) An 8-bit value (at the UNI and excludes the GFC field) used to identify an ATM path and carried in the cell header. Like a VCI, it is locally significant and refers to the VP active on the local UNI.

References

1. ITU-T G.711, "General Aspects of Digital Transmission Systems; Terminal Equipment Pulse Code Modulation (PCM) of Voice Frequencies," Geneva, 1972; further amended in 1988 and 1993.
2. ITU-T G.723, "General Aspects of Digital Transmission Systems; Dual Rate Speech Coder for Multimedia Communications Transmitting at 5.3 and 6.3 Kbps," 1996.
3. ITU-T G.726, "General Aspects of Digital Transmission Systems; Terminal Equipment 40, 32, 24, 16 Kbps Adaptive Differential Pulse Code Modulation (ADPCM)," Geneva, 1990.
4. ITU-T G.729, "General Aspects of Digital Transmission Systems, Coding of Speech at 8 Kbps Using Conjugate Structure Algebraic Code Excited Linear Prediction (CS-ACELP)."
5. ATM Forum af-pnni-0055.000, "Private Network-Network Interface Specification Version 1.0 (PNNI 1.0)," 1996.
6. ITU-T G.704, "General Aspects of Digital Transmission Systems, Transmission Systems Synchronous Frame Structures used at 1,544, 6,312, 2,048, 8,488 and 44,736 Kbps Hierarchical Levels."
7. ITU-T G.732, "General Aspects of Digital Transmission Systems Terminal Equipment, Characteristics of Primary PCM Multiplexing Equipment Operating at 2,048 Kbps."

8. DPNSS—BT Network Requirements Document—BTNR 188, BT Network Standards Secretariat, BT Angel Centre, 403 St. John Street, London EC1V 4PL.

9. ITU-T Q.931, "Digital Subscriber Signaling System No. 1 (DSS1)—ISDN User-Network Interface Layer 3 Specification for Basic Call Control." The complete volumes of the ITU material can be obtained from International Telecommunication Union, Sales and Marketing Division, Place des Nations – CH-1211, GENEVA 20 (Switzerland) Telephone +41 22 730 61 41 (English) / 41 22 730 61 42 (French) / +41 22 730 61 43 (Spanish) Telex: 421 000 uit ch / Fax +41 22 730 51 94 E-Mail:sales@itu.int / http://www.itu.int/publications

10. ITU-T Q.932, "Digital Subscriber Signaling System No. 1 (DSS1)—Generic Procedures for the Control of ISDN Supplementary Services."

11. ITU-T Q.933, "Digital Subscriber Signaling System No. 1 (DSS1)—Network Layer, D Signaling Specification for Frame Mode, Basic Call Control."

12. ITU-T Q.950, "Digital Subscriber Signaling System No. 1 (DSS1)—Stage 3 Description for Supplementary Services Using DSS1, Supplementary Services Protocols, Structure, and General Principles."

13. ITU-T Q.921, "Digital Subscriber Signaling System No. 1, ISDN User-Network Interface Data Link Layer Specification," 1997.

14. ITU-T I.363.1, "B-ISDN ATM Adaptation Layer Specification: Type 1 AAL," 1996.

15. ITU-T I.363.2, "B-ISDN ATM Adaptation Layer Specification: Type 2 AAL," 1997.

16. ITU-T I.366.1, "Segmentation and Reassembly Service Specific Convergence Sublayer for the AAL Type 2," 1997.

17. ITU-T I.366.2, "AAL Type 2 Service Specific Convergence Sublayer for Trunking," 1999.

18. ITU-T I.363.5, "B-ISDN ATM Adaptation Layer Specification: Type 5 AAL," 1996.

19. FRF.5, "Frame Relay/ATM PVC, Network Interworking Implementation Agreement," Frame Relay Forum, Suite 307, 39355 California Street, Fremont, CA 94538.

20. FRF.8, "Frame Relay/ATM PVC, Service Interworking Implementation Agreement," Frame Relay Forum, Suite 307, 39355 California Street, Fremont, CA 94538.

21. FRF.11, "Voice over Frame Relay Implementation Agreement," Frame Relay Forum, Suite 307, 39355 California Street, Fremont, CA 94538.

22. ITU-T G.114, "One-Way Transmission Time," 1996.

23. ATM Forum-97-1021, "Impact of Multiplexing and Buffering on VBR Voice Performance."

24. D. Anick, D. Mitra, and M.M. Sondhi, "Stochastic Theory of a Data-Handling System with Multiple Sources," *The Bell System Technical Journal* 61, no. 8 (October 1982).

25. T.H. Cheng, K.L. Cheah, E. Gunawan, and S.H. Oh, "A Moment Matching Approach for Modeling ATM Cell Arrival Processes," *International Journal of Communication Systems* 9, 1–3 (1996). 1996 © John Wiley & Sons Limited. Reproduced with permission.

26. John P. Cosmas, Guido Pete, Ralf Lehnert, Chris Blondia, Kimon Kontovassilis, Olga Casais, and Thomas Theimer, "A Review of Voice, Data, and Video Traffic Models for ATM," *European Transactions Telecommunications* 5, no. 2 (1994).

27. J. Daigle and J.D. Langford, "Models for Analysis of Packet Voice Communications Systems," *IEEE Journal on Selected Areas of Communication* SAC-4, no. 6 (September 1986).

28. C.J. Weinstein, "Fractional Speech Loss and Talker Activity Model for TASI and for Packet-Switched Speech," *IEEE Transactions on Communications* COM-26 (September 1978): 1,253–1,257.

29. Bellcore GR-1110-CORE Issue 1, "Broadband Switching System (BSS) Generic Requirement," September 1994.

30. P.T. Brady, "A Technique for Investigating On-Off Patterns of Speech."

31. ———. "A Model for Generating On-Off Patterns in 16 Conversations," *The Bell System Technical Journal* (January 1968): 73–91.

32. C.S. Chang, "Stability, Queue Length, and Delay of Deterministic and Stochastic Queuing Networks," *IEEE Trans. Automat. Contr.* 39 (1994): 913–931.

33. C.S. Chang, P. Heidleberger, S. Juneja, and P. Shahabuddin, "Effective Bandwidth and Fast Simulation of ATM in Tree Networks," *Perf. Eval.* 20 (1994): 45–66.

34. C. Courcoubetis, G. Kesidis, A. Ridder, J. Walrand, and R.R. Weber, "Call Acceptance and Routing in ATM Networks Using Inferences from Measured Buffer Occupancy," *IEEE Transactions on Communications* 43 (1995): 1,778–1,784.

35. S. Deng, "Traffic Characteristics of Packet Voice," *International Conference on Communications* no. 3 (January 1995): 1,369.

36. A.I. Elwalid and D. Mitra, "Effective Bandwidth of General Markovian Traffic Sources and Admission Control of High-Speed Networks," *IEEE/ACM Transactions on Networking* 1 (1993): 329–343.

37. P.W. Glynn and W. Whitt, "Logarithmic Asymptotics for Steady State Tail Probabilities in a Single-Server Queue," from J. Galambos and J. Gani, editors, "Studies in Applied Probability, Papers in Honor of Lajos Takacs," *Applied Probability Trust*, Sheffield, United Kingdom, 1994, 131–156.

38. R. Guerin, H. Ahmadi, and M. Naghshineh, "Equivalent Capacity and Its Application to Bandwidth Application in High-Speed Networks," *IEEE Journal on Selected Areas of Communication* 9 (1991): 968–981.

39. John G. Gruber, "A Comparison of Measured and Calculated Speech Temporal Parameters Relevant to Speech Activity Detection," *IEEE Transactions on Communications* COM-30, no. 4, (1982).

40. R.J. Gibbens, F.P. Kelly "Resource Pricing and the Evolution of Congestion Control," 1998, http://www.statslab.cam.ac.uk/~frank/evol.html.

41.———. "Distributed Connection Acceptance control for a connectionless network," June 1999, http://www.statslab.cam.ac.uk/~frank/evol.html.

42. G. Kesidis, J. Walrand, and C.S. Chang, "Effective Bandwidths for Multi-Class Markov Fluids and Other ATM Sources," *IEEE/ACM Transactions on Networking* 1 (1993): 424–428.

43. Harry Heffes and David M. Lucantoni, "A Markov Modulated Characterization of Packetized Voice and Data Traffic and

Related Statistical Multiplexer Performance," *IEEE Journal on Selected Areas in Communications* SAC-4, no. 6 (1986).

44. J.Y. Hui, "Resource Allocation for Broadband Networks," *IEEE Journal on Selected Areas in Communications* SAC-6 (1988): 1,598–1,608.
45. F.P. Kelly, "Effective Bandwidths at Multi-Class Queues," *Queuing Systems* 9 (1991): 5–16.
46. F.P. Kelly, Peter B. Key, and Stan Zachary, "Distributed Admission Control," Statistical Laboratory, University of Cambridge, Cambridge, CB2 ISB.
47. Leonard Kleinrock, *Queuing Systems Volume 1: Theory*, University of California, John Wiley & Sons, ISBN 0-471-49110-1.
48. ———. *Queuing Systems Volume 2: Computer Applications*, University of California, John Wiley & Sons, ISBN 0-471-49111-X.
49. M. Sexton and A. Reid, *Broadband Networking ATM, SDH, and SONET*, (Boston, Mass.: Artech House, 1997).
50. K. Sohraby, "On the Asymptotic Behavior of Heterogeneous Statistical Multiplexer with Applications," in IEEE INFOCOM 1992, Florence, Italy, 1992.
51. ———. "On the Theory of General On-Off Sources with Applications in High-Speed Networks," in IEEE INFOCOM 1993, San Francisco, Calif., 1993.
52. Sriram, Varshney, and Shanthikumar, "Discrete Time Analysis of Integrated Voice/Data Multiplexers with and without Speech Activity Detectors," *IEEE Journal on Selected Areas in Communications* SAC-4, no. 6 (September 1986).
53. W. Whitt, "Tail Probabilities with Statistical Multiplexing and Effective Bandwidths for Multi-Class Queues," *Telecommunication Systems* 2 (1993): 71–107.
54. Kotikalapudi, Sriram, and Whitt, "Characterizing Superposition Arrival Processes in Packet Multiplexers for Voice and Data IEEE Journal on Selected Areas in Communications SAC-4, no. 6 (September 1986).
55. Sriram, Lyons, and Yung-Terng, "Anomalies Due to Delay and Loss in AAL2 Packet Voice Systems," *IEEE Journal on Selected Areas in Communications* 17, no. 1 (January 1999).

56. Jingyu Qiu and Edward W. Knightly, "Measurement-Based Admission Control with Aggregate Traffic Envelopes," *IEEE/ACM Transactions on Networking* 9, no. 2 (April 2001).

57. ATM Forum/BTD-TM-01.02, "ATM Forum Traffic Management Working Group Baseline Text Document," Portland, July 1998.

58. ATM Forum af-tm-0056.000, "ATM Forum Technical Committee Traffic Management Specification Version 4.0," 1996.

59. ATM Forum af-tm-0121, "ATM Forum Technical Committee Traffic Management Specification Version 4.1," March 1999.

60. ATM Forum af-tm-0150, "Addendum to TM 4.1 for an Optional Minimum Desired Cell Rate for UBR," July 2000.

61. ATM Forum af-sig-0061.000, "ATM Forum Technical Committee User-Network Interface (UNI) Signaling Specification Version 4.0," 1996.

62. F. Allard, "Broadband Virtual Private Network Signaling," *BT Technology Journal* 16, no. 2 (April 1998).

63. James Roberts, Ugo Mocci, and Jorma Virtamo, "Broadband Network Teletraffic Performance Evaluation and Design of Broadband Multiservice Networks," Final Report of Action, COST 242, Springer Verlag, 1996.

64. ITU-T I.356 Series I, "Integrated Services Digital Network Overall Network Aspects and Functions—Performance Objectives B-ISDN ATM Layer Cell Transfer Performance," October 1996.

65. ITU-T I.371 Series I, "Integrated Services Digital Network—Overall Network Aspects and Functions—Traffic Control and Congestion Control in B-ISDN," August 1996.

66. ATM Forum/LTD-TM-01.09, Traffic Management Working Group Living List.

67. ATM Forum Technical Committee Traffic Management Working Group, Portland, July 1998.

68. Arnold O. Allen, *Probability, Statistics, and Queueing Theory with Computer Science Applications* (Academic Press, 1978).

69. Leonard Kleinrock, *Queuing Systems Volume I and II*, (New York: John Wiley and Sons, 1976).

70. Broadband Network Teletraffic—Performance Evaluation and Design of Broadband.

71. Roberts, Mocci, and Virtamo, "Multiservice Networks," Final Report of Action, COST 242, Springer.

72. Sriram, Lyons, and Yung-Terng, "Anomalies Due to Delay and Loss in AAL2 Packet Voice Systems: Performance and Methods of Mitigation," *IEEE Journal on Selected Areas in Communications* 17, no. 1 (January 1999).

73. D. Raychaudhuri and N. Wilson, "ATM-Based Transport Architecture for Multiservices Wireless Personal Communications Networks," *IEEE Journal on Selected Areas in Communications* 12 (October 1994): 1,401–1,414.

74. H. Nakamura, H. Tsuboya, M. Nakano, and A. Nakajima, "Applying ATM to Mobile Infrastructure Networks," *IEEE Communications Magazine* 36, no. 1 (January 1998): 66–73.

75. The ATM Forum af-vtoa-0083.000, "Technical Committee Voice and Telephony over ATM to the Desktop Specification," May 1997.

76. The Yankee Group, "Voice over Whatever: Internet, IP, Frame Relay, and ATM Find Their Voice Telecommunications Planning Service," December 1997, The Yankee Group, 31 St. James Avenue, Boston, MA 02116.

77. The Yankee Group, "Global Network Strategies Survey 2001: Report" 2, no. 13 (October 2001) by Sandra Palumbo, The Yankee Group, 31 St. James Avenue, Boston, MA 02116.

78. R.J. Gibbens and P.J. Hunt, "Effective Bandwidths for Multi-Type UAS Channel," *Queuing Systems* 9, no. 1 (1991): 17–28.

79. R.P. Swale, "VoIP—Panacea or PIG's Ear," *BT Technol Journal* 19, no. 2 (April 2001).

Bibliography

1. D. Ginsburg, *ATM Solutions for Enterprise Internetworking*, second edition, (Reading, Mass.: Addison-Wesley, 1999).
2. M. Sexton and A. Reid, *Transmission Networking: SONET and the Synchronous Digital Hierarchy*, (Boston, Mass.: Artech House, 1992).
3. Y. Bernet et al., "A Framework for Differential Services," Internet Draft, draft-ietf-diffserv-framework-00-txt, 1998.
4. D.V. Black et al., "An Architecture for Differentiated Services," Internet Draft, draft-ietf-diffserv-arch-00-txt, 1998.
5. M. Borden et al., "Integration of Real-time Services in an IP-ATM Network Architecture," RFC 1821, 1995.
6. ANSI T1.602-1996, "Integrated Services Digital Network (ISDN)—Data-Link Layer Signaling Specification for Application at the User-Network Interface."
7. ANSI T1.607-1990 (R 1995) and ANSI T1.607a-1996, "Digital Subscriber Signaling System Number 1 (DSS1)—Layer 3 Signaling Specification for Circuit-Switched Bearer Services."
8. ATM Forum af-ilmi-0065.000, "Integrated Local Management Interface (ILMI) Specification Version 4.0," 1996.
9. ATM Forum af-vtoa-0113.000, "ATM Trunking Using AAL2 for Narrowband Services," 1999.
10. ATM Forum af-cs-0125.000, "ATM Inter-Network Interface (AINI) Specification 1.0," 1999.

11. ETSI ETS 300 012-1, "Integrated Services Digital Network (ISDN); Basic User-Network Interface; Layer 1 Specification and Test Principles," 1998.
12. ETSI EN 300 347-1, "V5.2 Interface for the Support of Access Network (AN) Part 1: V5.2 Interface Specification," 1999.
13. ETSI ETS 300 402-1, "Integrated Services Digital Network (ISDN); Digital Subscriber Signaling System No. 1 (DSS1) Protocol; Data link Layer; Part 1: General Aspects," 1995.
14. ETSI ETS 300 402-2, "Integrated Services Digital Network (ISDN); Digital Subscriber Signaling System No. 1 (DSS1) Protocol; Data Link Layer; Part 2: General Protocol Specification," November 1995.
15. IETF STD0002 (RFC 1700), "Assigned Numbers," 1994.
16. ITU-T G.991.2, draft, "Single-Pair High-Speed Digital Subscriber Line (SHDSL) Transceivers."
17. ITU-T G.992.1, "Asymmetric Digital Subscriber Line (ADSL) Transceivers," 1999.
18. ITU-T G.992.2, "Splitterless Asymmetric Digital Subscriber Line (ADSL) Transceivers," 1999.
19. ITU-T G.964, "V-Interfaces at the Digital Local Exchange (LE)—V5.1-Interface (Based on 2,048 Kbps) for the Support Of Access Network (AN)," 1994.
20. ITU-T G.965, "V-Interfaces at the Digital Local Exchange (LE)—V5.2-Interface (Based on 2,048 Kbps) for the Support Of Access Network (AN)," 1995.
21. ITU-T I.430, "Basic User-Network Interface—Layer 1 Specification," 1995.
22. ITU-T I.432.1, "B-ISDN User-Network—Physical Layer Specification: General Characteristics," 1999.
23. ITU-T I.432.2, "B-ISDN User-Network—Physical Layer Specification: 155 520 Kbps and 622 080 Kbps Operation," 1999.
24. ITU-T I.432.3, "B-ISDN User-Network—Physical Layer Specification: 1,544 Kbps and 2,048 Kbps Operation," 1999.
25. ITU-T I.432.4, "B-ISDN User-Network—Physical Layer Specification: 51 840 Kbps Operation," 1999.
26. ITU-T I.432.5, "B-ISDN User-Network—Physical Layer Specification: 25 600 Kbps Operation," 1997.

27. ITU-T I.610, "B-ISDN Operation and Maintenance Principles and Functions," 1999.
28. ITU-T Q.2931 (02/95), "Broadband Integrated Services Digital Network (B-ISDN)—Digital Subscriber Signaling System No. 2 (DSS 2)—User-Network Interface (UNI) Layer 3 Specification for Basic Call/Connection Control."
29. ITU-T V.8, "Procedures for Starting Sessions of Data Transmission over the General Switched Telephone Network," 1998.
30. ITU-T V.25, "Automatic Answering Equipment and General Procedures for Automatic Calling Equipment on the General Switched Telephone Network Including Procedures for Disabling of Echo Control Devices for Both Manually and Automatically Established Calls," 1996.
31. Telcordia Generic Requirements GR-303-CORE Issue 2, "Integrated Digital Loop Carrier System Generic Requirements, Objectives, and Interface," 1998.
32. ANSI T1.101-1994, "Telecommunications—Synchronization Interface Standard."
33. ANSI T1.508-1992, "Network Performance—Loss Plan for Evolving Digital Networks."
34. DSL Forum WT-043 Rev 0.5, "Requirements for Voice over DSL," 2000.
35. ITU-T G.131, "Control of Talker Echo," 1996.
36. ITU-T G.168, "Digital Network Echo Cancellers," 1997.
37. Telcordia Technical Reference TR-TSY-000008 Issue 2, "Digital Interface Between the SLC(r) 96 Digital Loop Carrier and a Local Digital Switch," 1987.
38. Telcordia Technical Reference TR-NWT-000057 Issue 2, "Functional Criteria for Digital Loop Carrier Systems," 1993.
39. Telcordia TR-NWT-000393, "Generic Requirements For ISDN Basic Access Digital Subscriber Lines," 1991.

Index

Symbols

A

B

C

D

E

F

G

H

I

L

M

About the Author

Juliet Bates Ph.D. is a Principal Consultant in the Professional Services Group, Broadband Network Division, Alcatel Telecom UK. This book is based on her research and experience in implementing carrier networks. Dr. Bates lives in Maidenhead, UK.